AF440662

Monitoring of Pregnancy and Prediction of Birth-date

Enzymatic and Ultrasonographic Methods

Monitoring of Pregnancy and Prediction of Birth-date

Enzymatic and Ultrasonographic Methods

Rudolf Klimek

with contributions from
Alicja Rzepecka, Marek Klimek and
Andrzej Michalski

The Parthenon Publishing Group

International Publishers in Medicine, Science & Technology

LONDON   *CASTERTON*   NEW YORK

Published in the UK by
The Parthenon Publishing Group Limited
Casterton Hall, Carnforth, Lancs, LA6 2LA, UK

Published in the USA by
The Parthenon Publishing Group Inc.
One Blue Hill Plaza, PO Box 1564
Pearl River, New York 10965, USA

British Library Cataloguing in Publication Data
Klimek, Rudolf
  Monitoring of Pregnancy and Prediction
  of Birth-date: Enzymatic and
  Ultrasonographic Methods
  I. Title
  618.2

ISBN 1-85070-525-9

2nd English revised edition, Cracow

Typeset by Martin Lister Publishing Services, Bolton-le-Sands,
Carnforth, Lancs.
Printed and bound by Bookcraft Ltd, Midsomer Norton, Bath

Contents

Preface

Each epoch in the development of medicine leaves behind some long-lasting diagnostic and therapeutic achievements. Clinical enzymology is just one of those from the biochemical epoch. Nowadays it would be difficult to imagine clinical medicine without the possibility of determining aminotransferases, dehydrogenases or peptidases. It is amazing that obstetricians still look at oxytocinase determination with such disinterest as it is the most 'obstetric' enzyme, indicating the condition of the fetus, the placenta and the mother. It can be used for monitoring the course of pregnancy, estimation of a gestational age and expected time of delivery; for prediction of intrauterine fetal distress and death; probable occurrence of prenatal and perinatal hemorrhagic disorders and eventually for monitoring the efficacy of pharmacological and surgical treatment.

The aim of this book is to recapitulate 30 years of clinical experience in enzymatic monitoring of pregnancy and prediction of birth-date, which in the last decade included ultrasonographic measurements. Our biological-gestational scale, introduced to evaluate pregnancy development, comprises and links both enzymatic and ultrasonographic data, and in future will include other imaging techniques.

In modern obstetrics less technological advances such as electronic fetal monitoring, sonography or home uterine monitoring have come along too rapidly. There has not been enough time to prove new technology as being better than the established methods and rules, or simply to read and consider the older literature.

The present clinical view, that the most effective way to date pregnancy is by ultrasound alone, unfortunately has had a profound negative effect on the gestational age distribution by increasing the number of obstetric interventions and premature deliveries. A meta-analysis of many papers has shown no decrease in perinatal mortality or serious morbidity associated with preterm delivery, which is a problem throughout the world.

Fortunately, the constant increase in oxytocinasemia, just like sonographic measurements, can prove the child's appropriate development with all the more accuracy the more similar the subsequent values of the enzymatic and ultrasonographic determinations. This means that ultrasonographic fetometry, not only during the first and second trimesters but also within the third trimester, can be used to date the expected time of confinement and to predict fetal birth-weight.

Rudolf Klimek, M.D., Ph.D.
Professor and Chairman,
Univ. OB/GYN Institute,
31 – 049 Cracow, Poland

1
Introduction

Rudolf Klimek

Oxytocinase is an enzyme that decomposes the hypothalamic hormones. Because of this property, it belongs to the basic mechanisms of the human neurohormonal regulation system and is best known as the enzyme in the oxytocin–oxytocinase system. The most important feature of oxytocinase has proved to be the continuous increase in its concentration in the blood of pregnant women as their pregnancy advances. This shows a statistically significant correlation with increasing fetal and placental weight. Oxytocinase is produced in the placenta under the influence of endogenous and exogenous oxytocin and differs from its other isoenzymes in its optimal activity at pH 7.9. The tissue isoenzyme has maximal activity at pH of 6.2 and can be found in all tissues, regardless of pregnancy.

Oxytocinase has been finally characterized as an aminopeptidase, an enzyme that decomposes not only the cyclopeptide hormones with free cysteine amine groups present in their particles but also other simple aminopeptides, not only consisting of aminoacids. Oxytocinase, for example, frees β-naphthylamine, combined with L-leucine or L-cystine. Due to its ability to decompose L-cystine-di-naphthylamide it is also known as the cystineaminopeptidase, CAP, with the isoenzyme of the placental origin described as P-CAP (placental-CAP) or simply CAP_1, to differentiate it from the T-CAP (tissue-CAP) or CAP_2.

Initially, oxytocinase activity was determined by biological methods which investigated the disappearance of oxytocin or antidiuretic (vasopressin) properties. However, despite their high sensitivity, the results did not prove to be clinically repeatable. It was only after synthetic oxytocin and its analogs had been obtained and, after synthetic aminopeptides had been introduced, that more chemically stable and reproducible determination of oxytocinase was made possible. This happened 30 years ago, after a few years of clinical use of synthetic oxytocin and vasopressin, when there was a need to change and modify existing ideas about the time of pregnancy duration in order to be able to utilize oxytocin doses in a very precise chemically determined manner[1–5]. The observed cases of unexpected intrauterine fetal death or of uterine rupture could no longer be explained by the non-stabilized doses of natural hormones of the posterior lobe of the hypophysis – as it used to be called. However, we state with regret that even today, the modern pulsatory dosing with synthetic oxytocin or desaminooxytocin during automatically controlled infusions is still too often based upon incorrect assumptions, predating the era of enzymology and not always helpful in the proper induction of labor. Also, as shown in Chapter 7, enzymatic experience led to the proper interpretation of the ultrasonographic results. So it is worthwhile to reflect upon the sense of biological and calendar pregnancy duration, in order to improve understanding of the biochemical diagnosis and to make full use of the clinical possibilities provided by pregnancy monitoring by oxytocinase determination and ultrasonography.

A constant increase in blood oxytocinase level, just as in sonographic measurements, provides proof of the appropriate development of the fetus. Two consecutive values, measured at least 2 weeks apart, in an advanced pregnancy enable prediction of fetal birth-weight and birth-date within the whole range of birth occurrence, until 44 calendar gestational weeks; the more similar the subsequent values of enzymatic and ultrasonographic determinations, all the more accurate our predictions. Computerized programs automatically differentiate fast, regular and slow growing fetuses as well as psychological and pathological development of the individual pregnancy based on ultrasonographic data and gestational oxytocinase profiles. They indicate also biological gestational age without the need to use the date of the last menstrual period[6–13].

The authors are still using cross-sectional methods or scales of calendar gestational age assessment in pregnant women between 12 and 44 weeks' gestation. In consequence, the confidence intervals for any fetal ultrasound measurement increases with advancing gestation.

As a final result, the dependence on a single measurement to assess gestational age in the third trimester has a 95% confidence interval of ±3 weeks or more, but it could not be better than in nature. Contemporary ultrasonographers have only confirmed that the mean and the standard deviation of the duration of human pregnancy is 281±11 days. This was without doubt documented before the clinical application of ultrasound as well as being known previously from Aristotle's statement: 'All creatures have their determined time for giving birth and carrying fetus, only man is born all year long, not in determined time, one in the seventh month, the other in the eighth, and so on till the beginning of the eleventh month'.

2
Biological and calendar gestational age

Rudolf Klimek

The notion of time can mean a measurement analogous to the measurement of length, mass, speed, acceleration, etc. or it can mean the dimension of a phenomenon.

TIME AS A MEASUREMENT

Making use of time as a measurement we apply scales with their strictly defined units, e.g. seconds, hours, days, weeks, months or years. For example, epidemiological studies covering hundreds of thousands of pregnancies have shown that the average length of human pregnancy is 281 days counting from the first day of the last menstrual period (LMP). This is equal to 40 weeks and one day or 10 lunar months and one day[1-4,9,14-23].

It is also known statistically that about 66% of deliveries take place within the period of at least eleven days (lowest standard deviation in a normal human population) before and after the average gestational length. The range of two standard

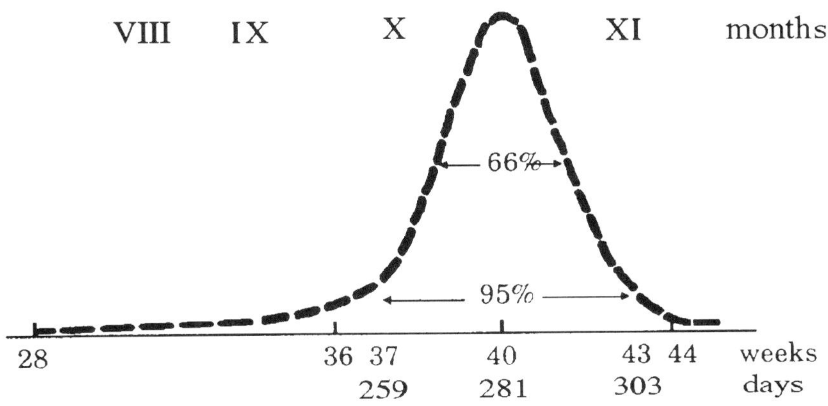

Figure 2.1 Normal distribution of length of human gestation (from R. Klimek[11])

deviations from the mean value is considered to be normal from the biological point of view and using the percentile scale, this should cover 95% of a given population. So, if we take into account 1000 pregnant women who had their LMP on the same day, 950 of them will deliver within the period 259–303 days (Figure 2.1).

On a scale of weeks this would mean that the range for the period of normal labor begins at $37^{0/7}$ weeks of pregnancy and ends at $43^{2/7}$ week counting from the first day of the last menstrual period. However, using a scale of lunar months, i.e. periods of 28 days, we may say that approximately 950 of 1000 women will deliver healthy babies within the period of the last three weeks of the tenth lunar month and the first three weeks of the eleventh month (Figure 2.2). It is necessary to state here that 5% of the babies (i.e. 50 cases) will be born healthy either just before the beginning of the physiologically normal period or just after its end.

On the other hand, counting the pregnancy duration from the moment of conception we could use the notion of real duration of pregnancy, by counting it 14 days earlier on a calendar scale. The trouble is that ovulation in healthy women with regular menstrual cycles occurs usually 14 days before the next menstrual period which would have appeared but for the pregnancy. Even in women with a normal menstrual cycle, which is notoriously variable within the range of several days, it would be difficult to use the real pregnancy duration, called postconceptional gestational age. Therefore, in clinical practice we use the notion of postmenstrual gestational time, age or period.

It is evident from the above considerations that the calendar scale of pregnancy length constitutes the time measurement of fetal development, like the measurements in kilograms of its body weight or in centimeters of the mother's abdominal circumference, etc.

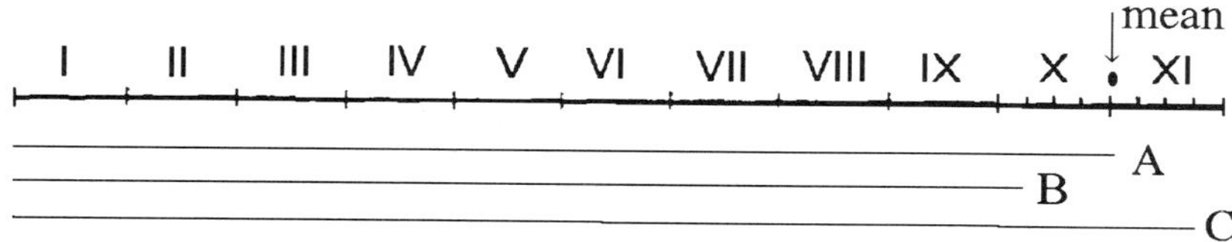

Figure 2.2 Postmenstrual gestational age in lunar months

TIME AS A DIMENSION

Disregarding the calendar meaning of time as a measurement, we should bear in mind that time is one of the most important dimensions of any living process. The proper understanding of the notion of dimension may be accurately illustrated using as an example the process of fetal maturation in the maternal womb, i.e. reaching the moment of being able to live independently outside the maternal body.

By convention, it is accepted that certain parameters of the fetus are regarded as criteria of its maturation, for example 2500 g of body weight is considered the lower normal limit of the ability to sustain independent life. We will use the mean body mass and the mean body length of a newborn as the criterion of average fetal maturity. Therefore, a mean body mass of 3400 g and mean body length of 54 cm, determine the moment at which the fetus reaches maturity in the mother's womb. However, one such baby, let us call it A, may be born as early as the 259th day of pregnancy, another (B) on the 281st day of pregnancy, a third baby (C) on the 288th day, a fourth baby (D) on the 295th day and a fifth baby (E) may be born towards the end of the normal biological range, on the 301st day of pregnancy (Figure 2.3).

It is obvious that all these infants (A–E) differ as far as their calendar age is concerned at birth, but are identical if their

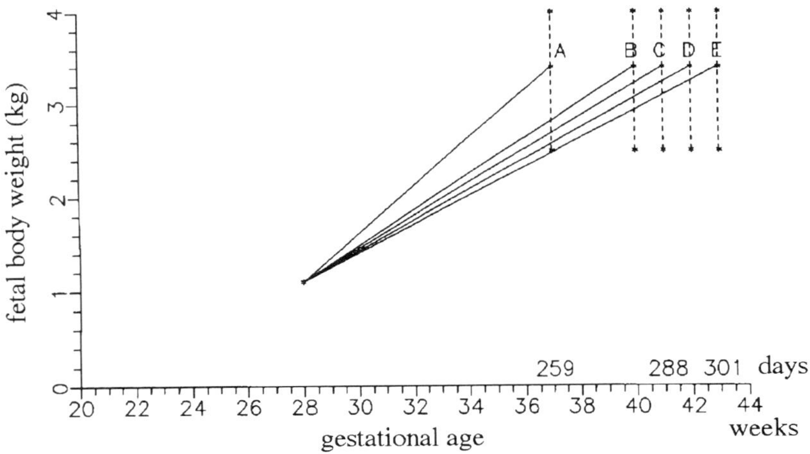

Figure 2.3 Mature biological age and its calendar measurements

biological age is taken into account, because all of them are mature enough to live outside the maternal organism.

The difference between the two infants: A and E on the calendar scale is as much as six weeks. It is well known that during such a long period of fetal development in the maternal womb, a baby is able to increase its body weight by 50%[11,19,24]. Baby A has already achieved 3400 g, while baby E by the 259th day of its calendar life has only achieved 2200 g of its body mass. Baby B on the same day of the calendar age scale weighs 2750 g. We have assumed that 3400 g constitutes the average criterion of maturity. It is obvious that baby E by the 259th day of its life is far from being mature, even in comparison with baby B which has already crossed the border of 2500 g, which is considered to be the lower normal limit for maturity.

The above examples illustrate in the best way the fact that biological age differs from calendar age, which carries extremely important practical consequences. In many countries all over the world, the rule widely applied for induction of labor is based exclusively on its calendar duration being longer than 293 days (so called post-term pregnancy). What is more, some other people try to move this border line to the seventh day after the mean duration of pregnancy marked by Naegelle's formula. Nevertheless, anybody who wants to induce labor at 294 or 295 days should bear in mind that such a procedure may be being undertaken sometimes too early. In the case of baby E the delivery is induced 7 or 8 days early (Figure 2.3). Whereas, if we consider the existence of fetuses that have already reached their maturity (e.g. A and B) and if spontaneous delivery has not started due to disturbances of the labor mechanisms, only waiting for day 294 or 295 to induce the delivery means a delay of several weeks which is not unimportant for the infant or the mother.

The principle of relativity of pregnancy duration, introduced by R. Klimek and originally related to the use of oxytocin in obstetrics, means practically that it is not the calendar time but the biological age that qualifies the results of both the diagnostic procedures and treatments[1–5].

Consideration of the biological state of the fetus was made possible by the introduction of the biochemical method for determination of the term of delivery, based upon the simultaneous use of the intravenous oxytocin test and determination of oxytocinase blood levels in the prepartal period (see pp. 30–34). Unfortunately, within the next few years the biochemical methods became dominated by the development of first cardiotocography and then ultrasonography.

TIME AS A FOURTH DIMENSION

It is due to ultrasonography that the fourth dimension, i.e. time, has been introduced for permanent medical use. The ultrasonographic technique enables us to examine the phenomena inside the maternal organism in three dimensions and to carry out the measurement of anatomical and functional development of the fetus, for example by measuring the blood flow in the fetal heart or the capacity of the auricles and ventricles of the fetal heart. This achievement has undoubtedly accelerated the development of NMR imaging of the gestational phenomena[12,25–27]. This new technique enables us to view in real time the beating heart of the baby inside the body of a completely dressed mother. What is more, simultaneously in selected sites of the heart muscle or directly in the flowing blood the concentration of chemical substances such as ATP, ADP, creatinine, lactic acid, etc. can be measured by NMR spectroscopy.

The observed images in three dimensions come in real time immediately from the resonating nuclei in the interior of the examined organisms. It is possible to distinguish, for example, the hydrogen in the $-CH_3$ group from that in the $=CH_2$ group and in the $\equiv CH$ group. These nuclei, by means of the radio-waves, can not only indicate their spatial localizations, but also inform to what degree they remain in balance within their own environment, obviously made up of homonymous elements or atoms. This can be done using the different times of NMR stimulation and registration of restored energy from the examined nuclei. It is possible to measure two different relaxation times, called longitudinal and transverse and therefore, to see different images, despite their reference to the same nuclei[177]. This means that besides the three-dimensional spatial localization: the length, width and depth as well as the fourth dimension – the real time, one and the same atom additionally can provide us with double information; so that while qualifying the state of health of a particular organism a fifth dimension is already taken into account.

On the atomic level, there is no possibility of the separation of matter from energy. Medicine has already reached the nuclear level. It investigates the phenomena on the level of atoms, nuclei and particles, where only the methods of statistical thermodynamics allow us to predict and interpret the phenomena that occur on the microscale. The phenomena we observe on the screen are the averaged images described by the general concepts of phenomenological thermodynamics.

Luckily, we do not have to dwell upon the description of biological phenomena in the 7th, 8th or 11th dimensions, used by contemporary physicists, mathematicians or some philosophers. We are satisfied with four dimensions, but in return we must be in agreement with the present state of knowledge, i.e. the examined phenomena will refer to biological time, the essential dimension of a child's development in the maternal womb.

By means of NMR we can observe the ovulation or nidation of the morula or the blastocyst in the gestationally changed endometrium. We can witness the existence of the new and unique biological unit, i.e. the zygote, the origin of human life, from the very moment of conception. The limit of purely biochemical diagnosis, for example gonadotrophin or progesterone measurements or the use of ultrasonography in order to determine the beginning of the preclinical or clinical pregnancy, has been shifted towards the earliest stage of pregnancy.

GESTATIONAL AGE

The calendar pregnancy length localizes the moment of conception on the daily, weekly or monthly calendar scale, only as the beginning of a process that progresses from the zygote to the mature stage achieved by a baby in the mother's womb. Biological pregnancy age designates not only the time needed for the full development of the baby but also the necessary time for complete maturation of the placenta and adaptation of the mother for the protection of fetal development and for giving birth. Therefore, biophysical or biochemical measurements, as well as evaluation of psychological states, have to be referred to the biological dimensions of the pregnancy and essential features of pregnancy and the infant's development, from conception until delivery.

The relativity of the gestational concept of time means that the biological, not calendar, time is the dimension determining the development of a pregnancy.

The calendar time makes up only one measure of the phenomena occurring in the pregnancy and enables us to localize them upon a daily, weekly or monthly scale. It is essential to maintain precision when dealing with the calendar duration of pregnancy and to report it only in the fully accomplished minutes, hours, days, weeks or months. The scale originates from the first day of the last menstrual period after which the woman became pregnant. The 259th day of calendar pregnancy means that the baby has lived for 259 days, all the same the pregnancy has lasted for 37 weeks. The age of a baby, living for 261 days, amounts to 372/7 weeks on the weekly scale. The average time of human pregnancy amounts to 281 days or 401/7 weeks. The 43 week-old infant has been living for 301 days, etc. If we want to designate the 303rd day on the weekly scale, the day that marks the end of the period of occurrence of normal births in human beings, we have to record it as 432/7 weeks.

Lack of proper differentiation of the above terminology leads to the paradox that has been officially propagated by the FIGO report[28]. It regards all preterm labors, meaning the ones that happen before the calendar norm of labor onset, as occurring before 37 weeks. This report recognizes the week figured on the scale of 400/7 to 406/7 week as the 40th week of the calendar pregnancy duration. But authors of this report made a basic mistake by regarding 42 (294 days or more) weeks already to be post-term gestational age. This statement, contrary to their proper definition of preterm gestational age, shortens the calendar norm by a whole 10 days. It limits the biological norm in practice too, by requiring special care or induction of labor in each post-term pregnancy instead of at the correct time, which is after the biological norm, i.e. after the completion of 432/7 weeks of pregnancy. In this way, apart from the daily, weekly and monthly scale of the calendar duration of pregnancy, the unique 'obstetric' scale of preterm, at-term and post-term has been introduced.

This pure obstetric scale introduces the faulty interpretation of biological phenomena as it suggests the necessity of intervention or lack of such a need due to calendar dates and not according to the biological age.

Figure 2.3 (see p. 7) shows that baby A, already mature on the 259th day of the calendar pregnancy duration, who was not born because of some disturbances in the labor inducing mechanism, has to be considered even at the beginning of the delivery norm to be a post-term baby from the biological point of view. The case of baby E, born physiologically at the end of the calendar norm is similar; being born 6 weeks later than case A cannot be considered to be a prolonged pregnancy in the biological sense.

Any baby that is born in the first half of the biological norm, which means in the period from 37⁰⁄7 to 40¹⁄7 weeks – can not only be considered immature but, in some cases, its pregnancy can be even regarded as biologically prolonged. And vice versa, the baby born on the 295th day, which means in the 42⁰⁄7 week, together with the babies that mature even later – until 43²⁄7 weeks, cannot be considered postmature but sometimes can be considered to be immature fetuses.

Therefore, exclusive calendar orientation means this total lack of concern for those infants, which due to maternal, individual or other reasons do suffer from developmental disturbances within the dimension of their biological age. It should be emphasized that the measurement of weight and length of the infant's body or its parts as well as tests correlated with those measurements performed inside the maternal organism or directly inside the fetus must always refer to its biological maturity as well as take into account its present biological age and the real gestational condition of the mother.

CLINICAL DEFINITION OF THE GESTATIONAL AGE

The normal range for the occurrence of delivery in 95% of cases is from 22 days before the mean which is the lower limit, to 22 days after the mean, the upper limit. Therefore, only 5% of cases will have normal deliveries outside of this biological range (Figure 2.4). The problem is determining in which part of the normal range any individual baby will be born.

From the clinical point of view 'term' as the expected date of birth, it is the time of full fetal maturation which may occur at any time within the whole normal range from 259 to 303 days! Therefore, if you would like to describe in which part of the calendar scale any fetus was born it should be defined as pre-range, in-range or post-range of normal occurrence.

One source of error has been the incorrect way in which doctors have divided the Gaussian curve into 'pre-' and 'post-term' periods, preterm being described as any baby born before 37 weeks, i.e. 3 weeks before the mean and post-term being described as any baby which is unborn 13 days after the mean duration of pregnancy, i.e. 281+13 days. Nevertheless, the baby called post-term may still have 10 days of the normal range left. This means that the 'obstetric' scale has no sense from a statistical point of view because the duration of the period

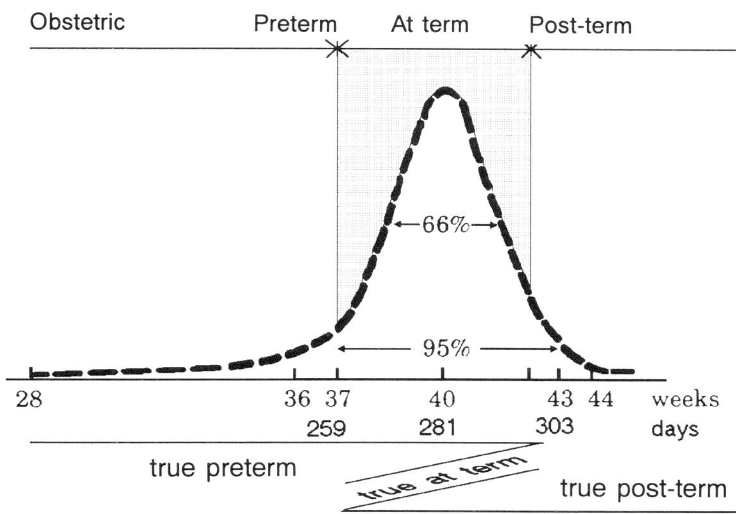

Figure 2.4 Alternative methods of defining the normal occurrence of delivery. The obstetric scale (top), division of Gaussian distribution of normal birth occurrence and its implication (bottom) (from R. Klimek[11])

called 'at term' has a range of 2 SD (22 days) from one side of the mean but less than 2 SD (only 13 days) from the other.

Term means the exact date of delivery in a normal case. Using such a definition of an individual term you can say that all babies who are born before the normal range (i.e. pre-range) are by definition preterm and all babies in which pregnancy is longer than the normal range (i.e. post-range) are post-term. However in addition, true pre- and post-term can also be used to describe some babies within the normal range for delivery.

For example, a baby who reached maturity at the beginning of the normal range, e.g. after 265 days, is true post-term if it is still not born by 270 days. Such a definition is also important in the case of induction of labor due to pregnancy lasting 10–14 days after the mean length of human gestation. If a baby was not due to reach maturity until the end of the normal range, by inducing labor too early the birth of a true preterm baby is being provoked.

It is very important, therefore, to look for the normal development of the pregnancy, to have well documented monitoring of the pregnancy, and to watch for the well-being of the baby, especially when we know that the maturation of the baby can be achieved within a range of 6 weeks. We also need to be very careful when describing the point of maturation of the fetus and the onset of labor whether spontaneous or provoked. Although in healthy women the natural process of birth occurs spontaneously in most cases, in some women pregnancy is truly prolonged because the mechanisms initiating labor are disturbed for different reasons.

In conclusion, from medical and ethical points of view we have to be opposed to the 'obstetric scale' comprising sequential periods: pre-, at- and post-term, because it is an incorrect way of calculating the duration of pregnancy. Obstetricians have to use 'term' only to describe an individual's date of fetal maturity within the normal range. We also state that the possible limits of pre- and post-term of an individual should be placed within as well as outside the normal range of deliveries.

In addition, all results given by ultrasound machines encompass not only mean values, with an accuracy in days (e.g. 35 weeks 2 days), but also their standard deviations or ranges in weeks (e.g. 35 weeks 2 days ±2 weeks), which is overlooked by doctors. For example, what $35^{2/7}$ ±2 weeks signifies is that there is a 95% chance the age is between $33^{2/7}$ to $37^{2/7}$ weeks and that the most likely age (but only <5%) is $35^{2/7}$ weeks. This is particularly important in late gestation and may have psychological as well as legal implications.

3

Biochemical monitoring of pregnancy

Rudolf Klimek and Andrzej Michalski

Almost all of the graphical scales and mathematical expressions used to describe the fetus' body mass, those taking into account an average and its standard deviations or the percentile intervals below 5% and above 95% with an average of 50% percentile, undergo a strange curvature after the 37th week of the calendar time of pregnancy duration (Figure 3.1).

It is even assumed that it is best to determine ultrasono-graphically an expected time of delivery around the 20th week of pregnancy, when the distribution of results obtained, within a span of ± 2 weeks, agrees with the distribution obtained in calendar time. Even in the classical handbooks on ultrasono-graphy it is very unusual to pay attention to the natural distribution of human deliveries and fetal body masses within

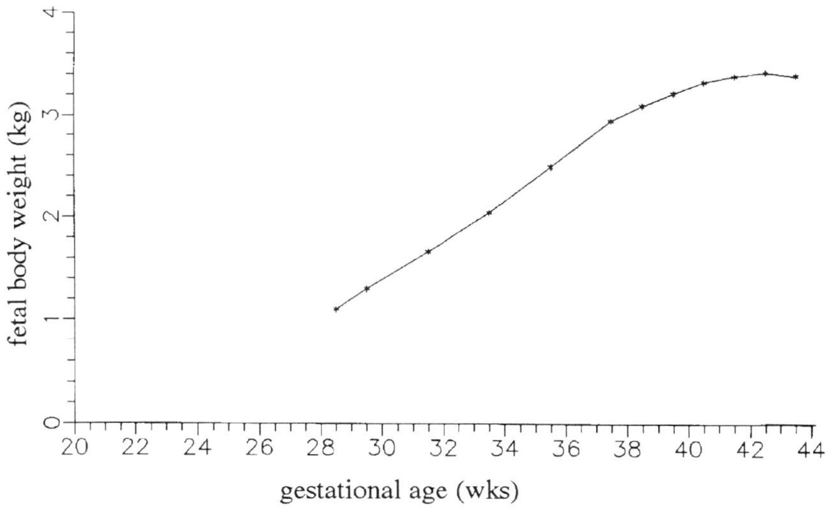

Figure 3.1 Fetal body weight throughout gestation measured by ultrasound

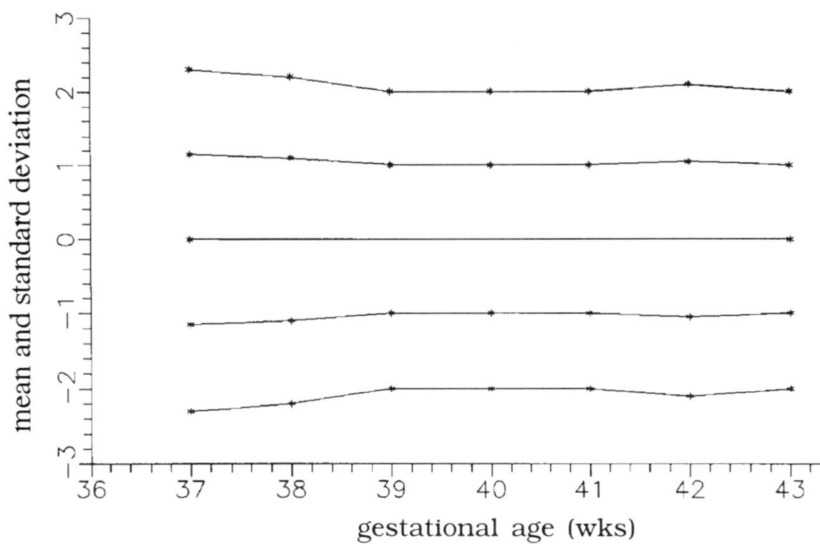

Figure 3.2 Mean ± standard deviations of birthweight-for-calendar gestational age

the calendar range of normal delivery occurrence (Figure 2.1, p. 5. The handbook written by Rudy Sabbagha which describes the distribution of human delivery occurrence to be that of a Gaussian curve, obtained on the grounds of 400 thousand documented cases, provides an exception[18].

However, if we measure and weigh mature babies born by normal labor during consecutive gestational weeks, it appears, on the basis of the previous and recent examinations, that the distribution of the birth weight in weekly intervals is practically identical (Figure 3.2). Our results obtained from observations in 1963 agree with those of other authors[16,29–32] and with new measurements of 1000 babies. In one-third of cases we were able to perform the enzymatic and ultrasonographic examinations simultaneously[6,33,34].

The divergence of the measurements results from reducing data from the whole biological norm to the average linear dependency instead of regarding it in the whole aspect (Figure 3.3). There is an attempt to reduce the natural distribution of infant's body mass (straight lines), to one average value, between the 37⁰/7 and 43²/7 week of pregnancy. In this way, the body mass of all babies, instead of only those with the same biological age throughout the whole period of natural labor occurrence, is resolved to one calendar age. The mass of babies to be born

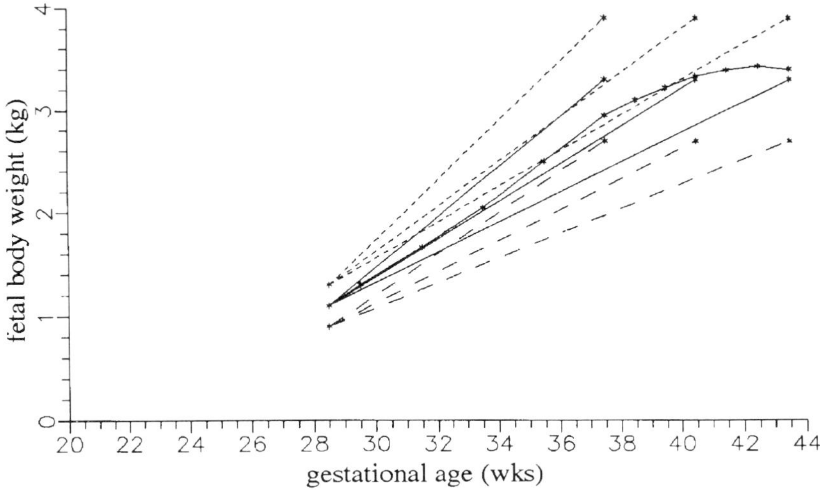

Figure 3.3 True fetal body weights (straight lines) and their ultasound average (curved line)

within a period of following weeks is added to the average mass, for example, of babies born at the beginning of the biological norm in the 37th week. Such methods finally lead to the false interpretation of the facts and build up false scales, referable to these measurements. Even less experienced doctors, measuring the particular parts of a fetus body, can easily distinguish between the fetus weighing 2100 g and that of 3300 g. This difference provides adequate distinction between

Table 3.1 Clinical data for calendar gestational age

Calendar age (wks)	>40	40	39	<39
Age years	26.4 ± 4.9	27.7 ± 5.2	28.1 ± 5.0	28.5 ± 5.7
Gravidity	1.7 ± 0.9	1.8 ± 1.0	1.9 ± 1.0	1.9 ± 1.2
Parity	1.5 ± 0.8	1.6 ± 0.7	1.6 ± 0.8	1.6 ± 0.9
Abortion	0.1 ± 0.4	0.2 ± 0.5	0.2 ± 0.5	0.2 ± 0.4
Children	0.5 ± 0.8	0.5 ± 0.8	0.6 ± 0.8	0.5 ± 0.8
Newborns				
weight kg	3.50 ± 0.5	3.53 ± 0.4	3.55 ± 0.4	3.48 ± 0.4
length cm	55.4 ± 3.2	55.1 ± 3.1	55.3 ± 2.9	54.9 ± 2.6
Apgar scores	9.7 ± 0.4	9.7 ± 0.5	9.7 ± 0.7	9.7 ± 0.6
Oxytocinase IU	8.3 ± 2.2	8.5 ± 2.0	8.3 ± 1.8	8.4 ± 2.2
N	89	128	166	258

N: number of oxytocinase measurements

the body masses of babies born at the beginning (37⁰/7) week and supposed to be born at the end (43²/7) of the normal calendar range for the occurrence of deliveries.

M. Klimek compiled (Table 3.1) data concerning the biological norm of birth occurrence (37⁰/7–43²/7). The results of those pregnancies delivered before the 39th week, in the 39th and 40th weeks and after the 40th calendar week, were separated into four groups.

The levels of prepartal oxytocinase (IU = μmol/l/min), determined $\leq$ 2 weeks before delivery, were identical. Also the inter-related measurements of the newborns and the obstetric history did not differentiate between these groups[6].

Lack of difference in the mean values of the above cases results from natural statistical tendencies that reveal themselves when dividing the data into one-week groups. The number of consecutive pregnancies decreases from 3.1± 1.2 in the 37th week to 1.2 ± 0.4 in the 43rd week on average, and the number of labors from 2.9 ± 0.9 to 1.2 ± 0.4 correspondingly, which relates to the decrease in the average number of previous children from 1.4 ± 0.7 to 0.2 ± 0.4. All labors began spontaneously in the 37th week, two-thirds of labors took place spontaneously between 38th and 41st week, in 42nd week only 40% and during the 43rd week there were no spontaneous labors only induced ones. The need for oxytocin use showed the opposite trend. Nevertheless, the weight and length of the newborns' bodies, and the Apgar scores were still independent of this considerably more detailed calendar scale. Therefore, regarding the mean fetal mass and its two standard deviations, the increase in mass should be applied not in terms of the

Table 3.2 Calendar gestational age and its ultrasound estimation ($\bar{x} \pm$ SD)

		Calendar age (weeks)		
Weeks	N	From LMP	From ultrasound	t
12–14	32	13.1 ± 0.8	12.8 ± 1.2	NS
15–18	43	16.2 ± 1.1	15.5 ± 2.8	NS
19–22	34	20.4 ± 1.1	19.9 ± 2.3	NS
23–26	30	24.9 ± 1.0	24.6 ± 1.4	NS
27–30	45	28.8 ± 1.2	28.9 ± 1.8	NS
31–34	50	32.5 ± 1.2	32.3 ± 1.9	NS
35–37	48	35.9 ± 0.8	36.0 ± 1.6	NS
38–	126	39.1 ± 0.9	37.6 ± 1.6	9.1***

N: number of determination; NS: not significant; *** $p < 0.001$

calendar but of biological gestational age.

Oxytocinase, together with its isoenzymes, reflects the biological pregnancy age, indicating the condition of the fetus and placenta as well as the mother. Other hormonal methods, evaluated most recently by A. Filipowicz in neuroendocrinologically successfully treated patients according to their infertility (Tables 3.2–3.5), do not show this convergence of enzymatically and ultrasonographically stated biological age. In particular estriol and HPL are not useful in advanced pregnancy assessment because of a considerable distribution of results[35].

In Table 3.2 calendar pregnancy age has been compared with the results of its ultrasonographic determination. Statistically significant differences in the comparable intervals of pregnancy were not noticed except after 37 weeks. The only difference was 1.5 weeks, within the norm of human delivery occurrence ($t = 9.1$, $p < 0.01$). This corresponds with data from the literature revealing the beginning of the important divergence occurring between the ultrasonographical determination of pregnancy duration and its calendar age only in the prepartal period[7, 36–58].

Table 3.3 Biological age of gestation for calendar age ($\bar{x} \pm$ SD)

Weeks to labor	N		From LMP	t	From ultrasound	t
					Calendar gestational age	
1	49		36.3 ± 1.4		37.3 ± 1.6	
				NS 1.3		**2.4
2–3	34	5.3 ***	36.8 ± 2.0		36.3 ± 2.0	
				***5.5		***4.6
4–6	31		34.0 ± 2.0		33.7 ± 2.4	
				***7.7		***6.3
7–10	52		30.4 ± 2.1		30.4 ± 2.2	
				***7.1		***7.5
11–14	37		26.3 ± 3.0		26.0 ± 3.0	
				***6.6		***6.1
15–18	33		22.0 ± 2.3		21.4 ± 3.2	
				***8.2		***5.6
19–22	36		17.5 ± 2.2		17.4 ± 2.6	
				***6.4		***6.1
23–26	38		14.8 ± 1.3		14.4 ± 1.4	
				***5.1		***4.8
27–30	12		12.7 ± 0.9		12.3 ± 0.9	

N: number of measurements; NS: not significant; ** $p < 0.01$, *** $p < 0.001$.

Table 3.4 Results ($\bar{x} \pm$ SD) of hormonal measurements

Weeks	N	HPL ng/ml	u	Estriol nmol/l	u
12–14	32	439.2 ± 317.7		102.1 ± 62.8	
			***4.9		***5.3
15–18	43	1271.3 ± 1047.3		205.5 ± 98.4	
			***4.8		***5.7
19–22	34	2210.8 ± 651.2		376.9 ± 159.1	
			***4.9		*2.0
23–26	30	3033.5 ± 695.6		807.9 ± 1197.8	
			***6.4		NS0
27–30	45	4436.2 ± 1262.3		802.5 ± 253.9	
			**2.7		*1.8
31–34	50	5356.7 ± 1884.3		969.8 ± 576.1	
			***3.4		*2.4
35–37	48	6715.8 ± 2006.3		1222.0 ± 505.4	
			NS1.9		*2.4
38–	126	6090.1 ± 1686.6		1434.7 ± 570.3	

Table 3.5 Results of hormonal measurements ($\bar{x} \pm$ SD) for biological gestational age*

Weeks	N	HPL ng/ml	u	Estriol nmol/l	u
1	49	6323.9 ± 1701.1		1441.3 ± 552.7	
			NS 0.3		NS 1.0
2–3	34	6441.4 ± 1885.6		1300.9 ± 649.3	
			NS 0.6		NS 0.8
4–6	31	6152.5 ± 2104.3		1167.1 ± 670.8	
			***3.9		**2.9
7–10	52	4531.6 ± 1290.1		805.7 ± 278.3	
			***3.6		NS 0.1
11–14	37	2580.3 ± 1200.0		824.7 ± 1078.1	
			***5.0		NS 1.8
15–18	33	2408.5 ± 741.4		470.2 ± 175.2	
			**3.1		***5.9
19–22	36	1666.3 ± 1196.9		250.5 ± 128.9	
			***4.0		***3.8
23–26	38	816.1 ± 458.8		152.4 ± 91.6	
			—		—
27–30	12	299.8 ± 265.8		76.2 ± 31.7	

N: number of measurements; *: lack of Student's test statistical differences; u: values of 'u' test.

Table 3.6 Biological gestational age for results of enzyme measurements ($\bar{x}$ ± SD)

Weeks to labor	N		Oxytocinase	*t*		Isooxytocinase	*t*
1	49		7.6 ± 1.7			6.4 ± 1.4	
				**2.7			*2.1
2–3	34	4.8 ***	6.5 ± 1.9		4.1 ***	6.7 ± 1.5	
				NS 1.1			NS 1.7
4–6	31		6.0 ± 1.5			5.1 ± 1.3	
				***4.8			***4.1
7–10	52		4.4 ± 1.4			4.0 ± 1.1	
				***4.5			***4.3
11–14	37		3.2 ± 0.9			3.1 ± 0.7	
				***6.9			***6.0
15–18	33		1.9 ± 0.6			2.2 ± 0.5	
				**2.7			*2.2
19–22	36		1.5 ± 0.6			1.9 ± 0.6	
				***3.3			NS 1.7
23–26	38		1.1 ± 0.4			1.7 ± 0.4	
				*2.3			***3.1
27–30	12		0.8 ± 0.3			1.3 ± 0.3	

N: number of measurements; NS: not significant; * $p < 0.05$, ** $p < 0.01$, *** $p < 0.001$

Nevertheless, when we compare the standard deviations, they distinctly appear to be considerably higher in the ultrasonographic group. In other words, the ultrasonographic designation of the calendar pregnancy age proves coherent until the 38th week of calendar scale, counted from the first day of LMP. This information is very significant as it indicates that, beginning from the 37th week of pregnancy duration, the usefulness of the calendar time of pregnancy duration to predict the onset of labor, is clearly decreasing, while the ultrasonographical measurements of fetal mass and length still reveal statistically significant differences (Table 3.3).

This significant ultrasonographical usefulness is contrasted with the results obtained while attempting to monitor pregnancy by means of estriol or HPL determination (Tables 3.4. and 3.5).

Table 3.4 shows that between the 12th and 14th weeks of pregnancy and its end, the level of estriol increases 14 times, and HPL 17 times with the greatest increment observed within the 35th to 37th week of pregnancy duration. Absence of a clinically explicit, perceptible increase of both hormones in consecutive time intervals, caused by considerable standard deviations and daily fluctuations of up to 50%, hinders the possibility of using these methods to monitor each advanced pregnancy[35].

The results of oxytocinase and isooxytocinase determination are quite different (Table 3.6). Comparison of the mean oxytocinase and isooxytocinase levels shows statistically significant differences between the consecutive time intervals of determination of both enzymes. From between the 12th and 14th week of pregnancy to its end, oxytocinase increase by 8.1 times, whereas the isooxytocinase level increases by 5.1 times. These increases in the mean values are thus two times lower than the estriol and HPL but predominate over the hormone determinations by exhibiting clinical and statistical useful difference in the consecutive enzyme's concentration. The increments in the consecutive weekly intervals of enzymes are linear.

To sum up, it is possible to evaluate the biological gestational age by means of ultrasonography and oxytocinase and isooxytocinase determinations, but determination of estriol and HPL, especially in advanced pregnancy, is of no value in clinical practice.

4

The clinical verification of biochemical methods of oxytocinase evaluation

Andrzej Michalski and Rudolf Klimek

Comparison of the accuracy of our enzymatic monitoring of pregnancy development and the onset of labor has been made possible by 30 years of application of the rule of relative duration of pregnancy. We compared the results of 1000 examinations obtained at the beginning of the 1960s in women with physiologically normal pregnancies and labor courses[2], with the latest results obtained in 233 women who delivered their babies spontaneously. The values of oxytocinase in the periods of pregnancy observed were compared.

The groups of pregnant women under observation were not different as far as pregnancy duration was concerned. This time equalled 281.6 ± 11 days 30 years ago and 280.7 ± 10.5 days nowadays[34]. The newborn's body weights were practically unchanged (3350 ± 430 g and 3440 ± 480 g, respectively).

The oxytocinase blood levels determined by means of the same method, (R. Klimek's modification of H. Tuppy and H. Nesvadba, presented in Chapter 5) remain the same as the data ob-

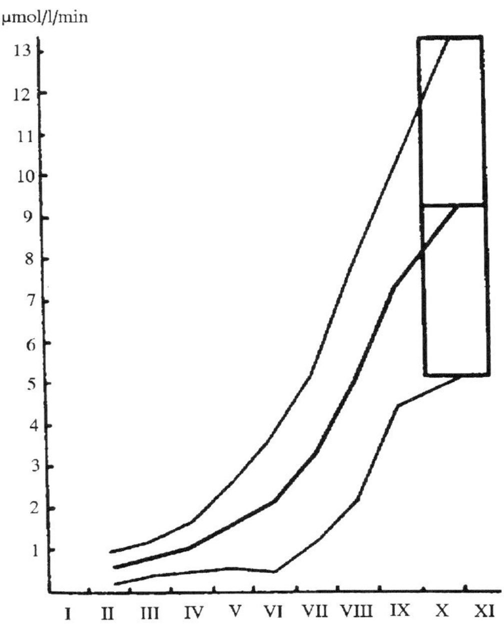

Figure 4.1 Oxytocinasemia in pregnancy ($\bar{x} \pm 2$ SD) (from R. Klimek[2])

[23]

tained 30 years ago (Figure 4.1) We can state that the applied method is characterized by considerable stability and repeatability. It permits the monitoring of the course of pregnancy by determining the biological time. When it was introduced, the results were compared with scientific reports at that time[59–64]. I. Nalepa compared it with five of the different available chemical methods, and it proved to be the best[65].

It is very important for every clinician to determine the repeatability of the method, both the absolute values as well as over the short periods of time. Three determinations performed at the beginning, in the middle and towards the end of pregnancy prove the considerable repeatability of results within consecutive weeks. If the values, in the prepartal period, have not increased or decreased by 0.2 IU (IU = μmol/l/min) labor occurrence will be soon. An increase of more than 0.2 IU means further pregnancy development within that week[6,10,33,66].

Our many years' experience have proved that, based upon the observation of the patients' clinical condition and the profile of the measured oxytocinase levels, one may suspect, and even prove, the existence of methodical errors. The results of oxytocinase determination have been compared between two centers, one that determines the oxytocinase only and the other that determines two isooxytocinases simultaneously. The only difference in the determination of the enzyme and its isoenzyme activity is the difference of optimal pH, results proved more repeatable in the center where both isoenzymes were determined (see p. 47).

OXYTOCINASE AND ISOOXYTOCINASE IN PHYSIOLOGICAL PREGNANCY

During physiological pregnancy, the oxytocinase level increases from the beginning of gestation until its end. This means that the maximum enzyme levels must occur in the normal range of labor occurrence in calendar time and more specifically in the last days before labor according to the biological age of pregnancy. Thus the highest values of enzyme levels, and therefore of the biological age of pregnancy, should be coherent with the results averaged for all values within the range of $37^0/7 - 43^2/7$ weeks and there can be no statistically significant differences between the two halves of this period ($37^0/7 - 39^6/7$ weeks and $40^0/7 - 43^2/7$ weeks). The contents of Table 4.1 illustrate this point[67].

In healthy women with a physiological course of pregnancy

Table 4.1 Results of enzyme measurements ($\bar{x} \pm$ SD, IU) for calendar and biological gestational age

	Enzyme			
	Oxytocinase	t	Isooxytocinase	t
Postmenstrual weeks				
40	8.3 ± 2.2		5.6 ± 0.9	
	(89)	NS	(5)	NS
40–37	8.4 ± 2.0		6.1 ± 1.4	
	(552)	***12.3	(71)	**2.8
36–33	6.7 ± 1.8		5.4 ± 1.5	
	(303)	***13.7	(69)	***4.1
32–29	4.8 ± 1.2		4.1 ± 1.8	
	(226)	***10.5	(48)	*2.0
28–25	3.5 ± 1.0		3.2 ± 1.8	
	(132)	***9.3	(26)	NS
24–21	2.4 ± 0.4		2.1 ± 0.1	
	(80)		(7)	
Weeks to labor				
0	8.2 ± 2.2		5.3 ± 1.4	
	(98)	NS	(14)	NS
1–2	8.3 ± 2.0		6.0 ± 1.2	
	(445)	**2.9	(53)	NS
3–4	7.8 ± 2.1		6.3 ± 1.7	
	(221)	***9.7	(40)	***5.2
5–8	6.1 ± 1.7		4.8 ± 1.0	
	(254)	***13.4	(52)	***6.6
9–12	4.3 ± 1.0		3.6 ± 0.8	
	(207)	***10.5	(50)	*2.4
13–16	3.1 ± 0.8		3.0 ± 0.6	
	(101)	***7.2	(12)	***3.1
17–20	2.3 ± 0.4		2.1 ± 0.2	
	(60)		(5)	

(N): number of measurements; NS: not significant; * $p<0.05$, ** $p<0.01$, *** $p<0.001$.

and spontaneous normal labor, the blood oxytocinase should reach its highest levels. On the contrary, in cases of induced labor, with delivery by Cesarean section there should be lower levels. These differences are evident (Table 4.2) and are all the more important as they are independent of the age of the healthy women and the number of previous children[67].

Similarly the isooxytocinase levels are lower in cases when it was necessary to assist the labor with oxytocin (Table 4.3).

Clinical observations only find their permanent place in

Table 4.2 Oxytocinase ($\bar{x} \pm$ SD µmol/l/min) for calendar and biological gestational age

Labor	Spontaneous			Induced		
Weeks	NL	Total	CS	Total	CS	NL
>40	9.1 ± 2.5 (36)	9.1 ± 2.5 (36)	—	7.7 ± 1.9 (53)	8.4 ± 1.2 (27)	7.0 ± 2.2 (26)
40–37	8.5 ± 1.9 (309)	8.6 ± 2.0 (360)	8.7 ± 2.2 (46)	8.0 ± 2.1 (192)	8.2 ± 2.1 (104)	7.7 ± 2.1 (88)
36–33	6.8 ± 1.8 (226)	6.9 ± 1.8 (260)	7.1 ± 2.1 (29)	5.7 ± 1.5 (43)	6.0 ± 1.6 (32)	5.0 ± 1.0 (11)
32–29	4.8 ± 1.2 (188)	4.8 ± 1.2 (214)	4.9 ± 1.3 (23)	4.6 ± 1.9 (12)	4.5 ± 2.0 (10)	5.3 ± 0.3 (2)
28–25	3.5 ± 0.8 (109)	3.5 ± 1.0 (121)	4.0 ± 2.0 (9)	3.2 ± 1.0 (11)	3.0 ± 0.9 (8)	3.5 ± 1.4 (3)
24–21	2.4 ± 0.4 (62)	2.4 ± 0.4 (70)	2.3 ± 0.6 (8)	2.1 ± 0.4 (10)	2.1 ± 0.3 (7)	1.9 ± 0.8 (3)
0	8.7 ± 2.2 (42)	8.6 ± 2.2 (52)	8.0 ± 1.8 (8)	7.7 ± 2.8 (46)	7.7 ± 1.9 (24)	7.8 ± 2.2 (22)
1–2	8.6 ± 1.9 (236)	8.6 ± 2.0 (276)	8.5 ± 2.2 (37)	7.9 ± 2.0 (169)	8.2 ± 1.8 (90)	7.6 ± 2.2 (79)
3–4	8.0 ± 2.0 (149)	7.9 ± 2.1 (175)	7.6 ± 2.5 (24)	7.3 ± 2.3 (46)	7.5 ± 2.7 (32)	6.8 ± 1.1 (14)
5–8	6.2 ± 1.6 (190)	6.2 ± 1.7 (218)	6.1 ± 2.3 (24)	5.5 ± 1.8 (36)	5.8 ± 2.0 (24)	5.0 ± 0.9 (12)
9–12	4.4 ± 1.1 (184)	4.3 ± 1.0 (203)	3.7 ± 0.8 (15)	3.7 ± 1.0 (4)	3.3 ± 0.3 (3)	5.2 ± 0.0 (1)
13–16	3.2 ± 0.7 (82)	3.2 ± 0.8 (88)	2.2 ± 0.2 (5)	2.9 ± 0.7 (13)	2.9 ± 0.8 (10)	2.7 ± 0.2 (3)
17–20	2.4 ± 0.4 (51)	2.4 ± 0.4 (53)	1.8 ± 0.5 (2)	1.7 ± 0.2 (7)	1.8 ± 0.2 (5)	1.5 ± 0.1 (2)

(N): number of enzyme measurements; NL: natural labor; CS: Cesarean section.

medical practice when observed phenomena have been stable for dozens of years and it is possible to reduce them to clear and explicit rules. This problem was discussed in one of our papers[68] in which the aim of the study was to evaluate the blood level of oxytocinase and isooxytocinase during the last 6 weeks of pregnancy according to the biological and calendar gestational age. The observations were carried out in 200 women whose pregnancies ended after $40^{1}/7 \pm 1.5$ weeks. Sonographic and enzymatic monitoring of these pregnancies was performed. The patients were divided into two groups: I – those who delivered spontaneously and II – those who delivered after

Table 4.3 Isooxytocinase ($\bar{x} \pm$ SD µmol/l/min) for calendar and biological age

Labor	Spontaneous			Induced		
Weeks	NL	Total	CS	Total	CS	NL
>40	5.4 ± 1.7 (2)	5.4 ± 1.7 (2)	—	5.7 ± 0.4 (3)	—	5.7 ± 0.4 (3)
40–37	6.3 ± 1.6 (26)	6.4 ± 1.7 (41)	6.9 ± 2.1 (12)	5.6 ± 0.6 (30)	5.2 ± 0.5 (12)	5.8 ± 0.5 (18)
36–33	5.6 ± 1.5 (45)	5.8 ± 1.5 (51)	7.1 ± 1.7 (6)	4.5 ± 0.6 (18)	4.2 ± 0.4 (14)	5.3 ± 0.4 (4)
32–29	4.0 ± 0.8 (43)	4.1 ± 0.8 (45)	5.2 ± 0.0 (2)	3.9 ± 0.4 (3)	3.9 ± 0.4 (3)	—
28–25	3.2 ± 0.1 (21)	3.3 ± 0.8 (24)	3.8 ± 2.2 (3)	2.6 ± 0.0 (2)	2.6 ± 0.0 (2)	—
24–21	2.1 ± 0.2 (5)	2.1 ± 0.1 (7)	2.0 ± 0.0 (2)	—	—	—
0	4.9 ± 0.8 (3)	5.7 ± 1.6 (7)	6.7 ± 2.2 (3)	5.0 ± 1.2 (7)	4.6 ± 1.0 (5)	6.1 ± 1.0 (2)
1–2	6.1 ± 1.1 (21)	6.2 ± 1.4 (33)	6.8 ± 1.8 (10)	5.7 ± 0.6 (20)	5.2 ± 0.4 (8)	6.0 ± 0.5 (12)
3–4	7.0 ± 1.8 (19)	7.1 ± 1.8 (25)	7.3 ± 2.0 (6)	5.1 ± 0.4 (15)	4.7 ± 0.3 (7)	5.5 ± 0.1 (8)
5–8	4.9 ± 1.1 (39)	5.0 ± 1.0 (41)	5.2 ± 0.0 (2)	4.3 ± 0.6 (11)	3.9 ± 0.3 (8)	5.1 ± 0.4 (3)
9–12	3.6 ± 0.8 (48)	3.6 ± 0.8 (49)	2.6 ± 0.0 (1)	4.4 ± 0.0 (1)	4.4 ± 0.0 (1)	—
13–16	3.4 ± 0.3 (7)	3.0 ± 0.7 (10)	2.1 ± 0.2 (3)	2.6 ± 0.0 (2)	2.6 ± 0.0 (2)	—
17–20	2.1 ± 0.2 (5)	2.1 ± 0.2 (5)	—	—	—	—

(N): number of enzyme measurements; NL: natural labor; CS: Cesarean section.

pharmacological induction of labor.

Table 4.4 shows the results of oxytocinase determination in the observed group of patients. It is clear that if we compare the mean values of blood oxytocinase levels in groups I and II we will see that in group I (spontaneous labor) mean oxytocinase levels are higher than in group II (induced labor) in all of the week intervals. The level of oxytocinase increases up till the day of delivery, unless induction of labor is necessary, when the level becomes stable from the 37th week of pregnancy on the calendar scale.

The isooxytocinase level is different (Table 4.5). Its values are also higher in spontaneous labor (group I), but in calendar

Table 4.4 Oxytocinasemia ($\bar{x} \pm$ SD µmol/l/min) in pregnant women

Weeks postmenstrual	40	39	38	37	36	35
Group I	9.2 ± 2.4	8.5 ± 1.7	8.5 ± 1.8	8.3 ± 1.9	7.9 ± 1.7	6.8 ± 1.7
Group II	7.0 ± 0.6	7.5 ± 1.1	7.4 ± 1.3	7.8 ± 1.6	5.4 ± 1.0	4.9 ± 0.8
Weeks to labor	<1	1	2	3	4	5
Group I	8.8 ± 2.0	8.6 ± 2.1	8.3 ± 2.0	7.4 ± 2.0	7.3 ± 1.9	6.4 ± 1.2
Group II	8.4±2.0	7.7±0.9	7.4±1.1	7.3±1.2	6.1±0.5	5.7±0.4

Table 4.5 Isooxytocinasemia ($\bar{x} \pm$ SD µmol/l/min) in pregnant women

Weeks postmenstrual	40	39	38	37	36	35
Group I	6.7 ± 1.9	5.5 ± 0.9	7.0 ± 1.9	6.8 ± 1.7	6.9 ± 1.4	4.8 ± 0.8
Group II	6.1 ± 0.6	5.9 ± 0.6	5.5 ± 0.2	5.5 ± 0.1	5.5 ± 0.2	4.6 ± 0.1
Weeks to labor	<1	1	2	3	4	5
Group I	6.4 ± 1.1	5.6 ± 0.9	7.5 ± 1.8	6.2 ± 1.5	5.9 ± 1.1	5.1 ± 0.8
Group II	5.9 ± 0.5	6.1 ± 0.5	5.8 ± 0.5	5.4 ± 0.2	5.5 ± 0.2	5.4 ± 02.

age they reach the plateau 5 weeks, and in biological age 3 weeks, before spontaneous labor. In induced labors isooxytocinase reaches a plateau later, about 2–3 weeks before labor.

This confirms a constant increase in oxytocinase until spontaneous labor and its higher level in spontaneous deliveries in comparison with pregnancies which end with induced labor. In relation to the above it is interesting to note different levels of isooxytocinase which reach a plateau 2–5 weeks before labor. What is more, this takes place later in the cases of induced delivery.

The results confirm the clinical applicability of enzymatic monitoring of pregnancy. The basic conclusion of this work is the following clinical indication: a constant increase of oxytocinase is evidence of proper development of pregnancy, and the plateauing of the level of isooxytocinase indicates a forthcoming labor.

OXYTOCIN–OXYTOCINASE SYSTEM

K. Semm explained the fact that the highest oxytocinase concentration in the blood occurs during the period of delivery by postulating a protective role of the enzyme against injury of the uterus when under maximum strain. Although it is difficult to agree that such a function be ascribed to the enzyme, its role in lowering the peak concentration of oxytocin (e.g. exogenous oxytocin given intravenously) is beyond doubt[69,70].

In R. Klimek's opinion the highest level of oxytocinasemia should be considered a result of the organism's potential enzyme synthesizing capacity; further induction is followed by a lasting shift of the oxytocin–oxytocinase system (O–O.S) equilibrium towards the hormone. In earlier stages of pregnancy the actual enzyme level does not correspond with the full adaptability of the organism. Under the influence of endo- or exogenous oxytocin the actual level of oxytocinasemia is temporarily increased by the organism[2-5,10,71,72]. This ability of the organism, which is one of the fundamental mechanisms supporting the pregnancy, disappears at the end of pregnancy as a result of lowered adaptability of the placenta, the main source of oxytocinase. At this time the O–O.S equilibrium can be altered and the labor activity of the uterus initiated even by minimal amounts of oxytocin. To obtain such an effect in the earlier stages of pregnancy, much higher doses of the hormone must be given for a longer period, partly to overcome the additionally increased oxytocinase activity.

The difference in oxytocinasemia before and after oxytocin administration (actual and potential oxytocinasemia) may be a measure of this enzymatic block. This difference is highest in the middle-period of pregnancy and reaches 4–5 IU of the enzyme. The lowest ones are observed in labor. Thus the actual oxytocinase level is a measure of the metabolic activity of endogenous oxytocin. The rational use of oxytocin in obstetrics demands an appreciation of the limitations of this induction as well as of the need to overcome the enzymatic block before clinical effects can be achieved.

DYNAMIC OXYTOCIN TEST

Since the first industrial synthesis of oxytocin[59], Syntocinon has proved the most essential and most frequently used preparation in modern obstetrics. However, the effects of identical doses may be different in women at the same stage of pregnancy. For this reason every estimation of the effect of a dose of the hormone on uterine contractility should be considered more important for further dosage than taking into account only the empirical international units. Moreover, before oxytocin is used therapeutically in the perinatal period, the term of pregnancy should be estimated by the biochemical method[3,4]. During the 44-day biological norm, optimal conditions for inducing labor certainly exist for at least several days. Therefore the earlier before this period these efforts are made to initiate delivery, the greater is the activity of pregnancy-protecting mechanisms, i.e. mechanical, enzymatic and progesterone blocks, to be overcome.

Biochemical estimation of term consists of investigating the condition of O–O.S. A lasting shift of its equilibrium towards the hormone determines the onset of labor. A continuous increase in oxytocinasemia indicates normal development of pregnancy, while a positive intravenous oxytocin test permits estimation of term. It must be appreciated that the Syntocinon test is based on the dynamic increase in the uterine sensitivity to oxytocin in the final phase of pregnancy. Therefore the test should be repeated several times, every other determination being made at the end of a period estimated by the preceding test.

The oxytocin test, even if dynamic, cannot be an independent safe clinical method because too frequently it yields identical results during a certain period. From the modern medical viewpoint, clinical tests of this kind should fulfil the following conditions:

(1) The results should be strictly dependent upon accurately measured doses;
(2) The manner in which the exogenous substances are administered should conform to physiological conditions or imitate them as closely as possible;
(3) The test must be reproducible and its evaluation must depend as little as possible on subjective data derived from the person performing the test and the subjects being tested; and

(4) The results should only be treated as an aid to clinical diagnosis and are not a substitute for other clinical criteria on which treatment depends.

Only the intravenous administration of oxytocin fulfils these conditions. It will suffice to mention the difficulties encountered when employing aerosols and sprays, the effects, for instance of inflammatory changes in the nasal and oral mucous membranes on hormone absorption, for it is always possible that inflammatory changes are present. Finally there is the presence of tissue oxytocinases and their effect on preparations administered intramuscularly or hypodermically.

C.N. Smyth's achievement is not only a suitable intravenous route, but primarily the demonstration that there is a quantitative relationship between uterine contractile sensitivity and the dose administered in his test[73]. This relationship was confirmed in an enormous number of experimental and clinical materials. The practical application of this was in agreement with Smyth's original directions for predicting the effectiveness of induction of labor. Test results with about 20 mU and also in a smaller percentage of reports with about 30 mU are regarded as a positive test result and are the basis for inducing labor.

Reports by various other authors are not so consistent in predicting the beginning of spontaneous labor on the basis of the well known table by C.N. Smyth. In particular, the too frequent positive results which can be obtained with the test over several weeks called into question the clinical usefulness of the oxytocin test. Double tests at intervals designed to increase the accuracy of results have also been of little help. However, the biochemical interpretation of the oxytocin test has brought a real change in its clinical usefulness[72,74–80]. This is summarized in Table 4.6.

A directly proportional relationship exists between the size of the doses of exogenous oxytocin necessary to induce the first uterine contraction and the length of the period between the test and the beginning of spontaneous labor. In the last week of pregnancy every 10 mU administered at 1 min intervals means one day to term. Although in more than 50% of cases this relationship is exact, in the others labor is considerably delayed. Because of this the oxytocin test should be repeated towards the end of the period indicated by the initial test and in the absence of clinical symptoms of uterine contractions or their intensification. In physiological conditions, the next test values are usually lower and in more than 80% of cases they

Table 4.6 Prediction of confinement according to R. Klimek or C.N. Smyth

Results of test ml (milliunits of oxytocin)	Predicted date after days	
	C.N. Smyth	R. Klimek
1 (10)	0	1
2 (20)	0–1	2
3 (30)	1	3
4 (40)	2	4
5 (50)		5
6 (60)	3–4	6
7 (70)		7
8 (80)	5–6	
9 (90)		
10 (100)	7	

indicate the term of delivery. The test performed in this way has been called dynamic by R. Klimek in contrast to the static test devised by C.N. Smyth.

It is sometimes necessary and possible to differentiate between the same test values by simultaneously measuring the oxytocinase blood levels. Needless to say, identical test values in the same women obtained in subsequent tests must be differently interpreted, if the oxytocinase blood levels measured at the same time are different. The correct obstetric evaluation of the Syntocinon test and of oxytocinasemia constitutes the complete biochemical method of estimating term of delivery and makes it possible to follow the progress of pregnancy.

It is possible to describe in a simple fashion the relationship between the values obtained in the oxytocin test and the oxytocinasemia, indicating the subsequent oxytocin test values by Roman numerals as follows: decrease as I, persistence of identical values as II and increase as III. Similarly an increase in enzyme concentration in the blood is indicated by the letter A, a consistent level by B and a decrease by C. Table 4.7 shows the combinations of results obtained in observations on several hundred cases[75,81].

Groups: IA, IB, IC, and IIA, IIIA relate to physiological conditions and account for 90.5% of the material. Group IIB (4.9%) represents a transition to other conditions in which, instead of passively waiting for labor to commence spontaneously, we induce labor. The introduction of the Syntocinon test as a biochemical method for following the course of pregnancy makes it possible to estimate term accurately to within

Table 4.7 Prepartal state of the oxytocin–oxytocinase system %

| | Oxytocin test | | | |
Oxytocinasemia	I decrease	II constant	III increase	Total
A increase	77%	1.5%	1.5%	80%
B constant	4.4%	4.9%	0.7%	10%
C decrease	6.1%	2.1%	1.8%	10%
Total	87.5%	8.5%	4%	100%

a few days in every case. The effectiveness in inducing labor is over 90%. Today, instead of the oxytocin test we can use the clinical achievements of cardiotocography.

Normally the subsequent levels of oxytocinasemia, as determined simultaneously with the oxytocin sensitivity tests, show an increase. Their persistence at a stable level, accompanied as a rule by stable test results, indicates that the normal labor mechanism is disturbed and that obstetrical intervention is necessary.

A fall of enzyme concentration without simultaneous contractions of the uterus indicates the imminent death of the fetus *in utero.*

Moreover, enzyme levels beyond the limit of 6 μmol/l/min are an indication that the range of normal body weight has been reached by the fetus. Together with a value of ≤30 mU obtained by the Syntocinon test it enables delivery of a mature fetus and labor to be successfully induced.

Therefore, an essential advantage of this method is that the physician can obtain two objective, reproducible and interrelated dynamic values of the oxytocin test and oxytocinasemia. Their proper obstetric interpretation constitutes what is referred to as the biochemical method for estimating the term of delivery. It was twice subjected to a clinical trial by K. Drewniak who found it to be ten times more accurate than Naegele's rule for estimating the term of labor in the individual pregnant woman[75,76]. The latter one remains of essential clinical importance in that it determines the first day of 11th lunar month of pregnancy, i.e. the middle day of the 44-day period in which physiological delivery occurs in human beings. However, it should be borne in mind when applying this rule that the date on which the last menstruation began is estimated by women with an error of several days; 3050 women in the last week of pregnancy were questioned as to the day of the month on which their last menstruation began[2]. From the statistical viewpoint, if the 31st day is discounted, 100 answers (3.3%) should fall on each day. The results clearly showed that pregnant women do not remember the date of the last menstruation (about 7% of them) or they estimate it incorrectly, either through fondness for certain numbers (e.g. 10, 15, 20) or for reasons known only to themselves. Of course this has an influence on our estimations of time intervals in pregnancy. Using computerized prediction of birth-date and birth-weight it is now not necessary to take into consideration LMP at all (see p. 53).

5
Oxytocinase monitoring of high-risk pregnancies

Rudolf Klimek and Marek Klimek

While investigating the level of oxytocinase and its isoenzymes, the possibility of predicting a high risk pregnancy's transition into the state of clinical threat, expressed for example by uterine contractions, spotting, bleeding and their accompanying subjective symptoms, seems most advantageous. Pregnancy's enzymatic monitoring is based upon the very well verified principle which comprises the rule that the constant increase of oxytocinase level until delivery proves the proper course of pregnancy, whereas a plateau in the increased isooxytocinase level signals the advent of labor (Figure 5.1).

Earlier than 2 weeks before delivery, reduction of this particular increase of oxytocinase level, especially its inhibition or decrease, testifies possible fetal distress. This rule has been

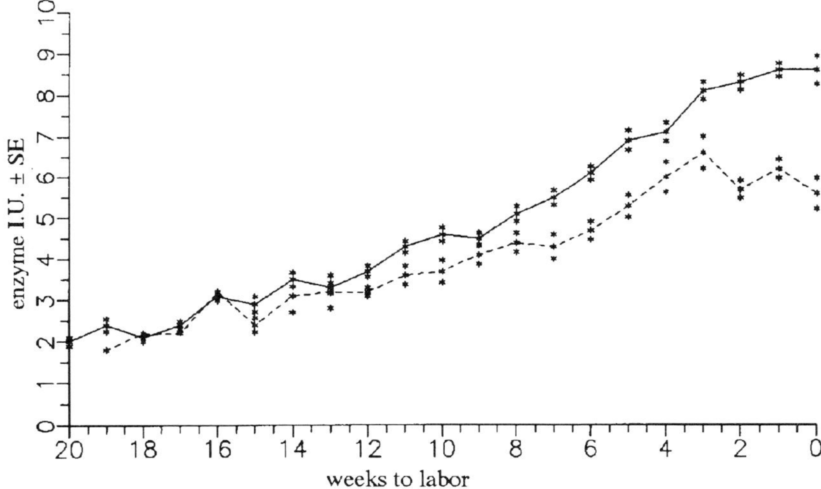

Figure 5.1 Gestational profiles of oxytocinasemia (—) and isooxytocinasemia (- -) in healthy pregnant women (mean ± standard error μmol/l/min) (from M. Klimek[6])

[35]

verified in pregnancies complicated by diabetes or hypertension, the ones that most often end with premature delivery.

DIABETES AND HYPERTENSION IN PREGNANCY

In their research, M. Krzyczkowska-Sendrakowska and co-authors studied a group of 107 diabetic women[82]. The group was divided according to A. White's classification: G, 27 cases; B, 35 cases; C, 25 cases; and RF, 20 cases. The average of their age was 28.8 ± 4.3 years, with an age range of 18 to 43 years. The pregnancy duration amounted to 38.3 ± 1.8 weeks, the newborns' body weight: 3270 ± 652 g; only in 17 (15.9%) cases was premature labor observed.

The second group comprised 50 women with hypertension, an average age of 32.4 ± 3.7 years, and an age range of 21 to 42 years. There were 17 women who suffered from hypertension before conception; in 33 cases pregnancy-induced hypertension was diagnosed. The pregnancy lasted for 37.9 ± 2.7 weeks on average, the newborns' body weight was 2738 ± 863 g and the number of premature labors was 13 (26%).

The study groups have undergone additional division into three sub-groups: A, with a stable increase in oxytocinase level; B, with final maintenance or decrease of oxytocinase level

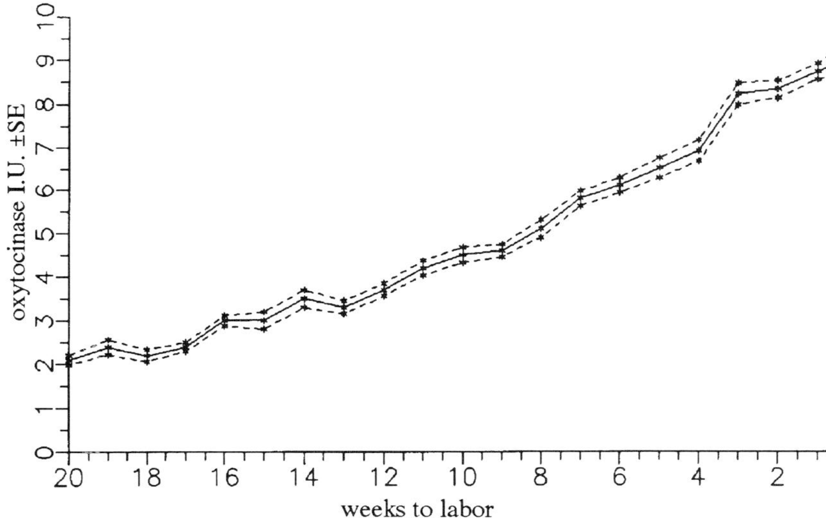

Figure 5.2 Mean values and their standard error of increasing profile of oxytocinasemia in healthy pregnant women (from M. Klimek[6])

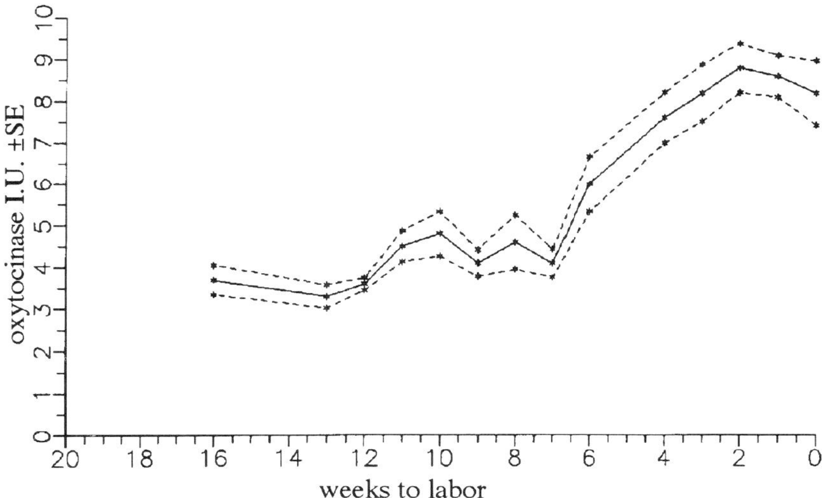

Figure 5.3 Mean values and their standard error of prepartal decreasing profile of oxytocinasemia in healthy pregnant women (from M. Klimek[6])

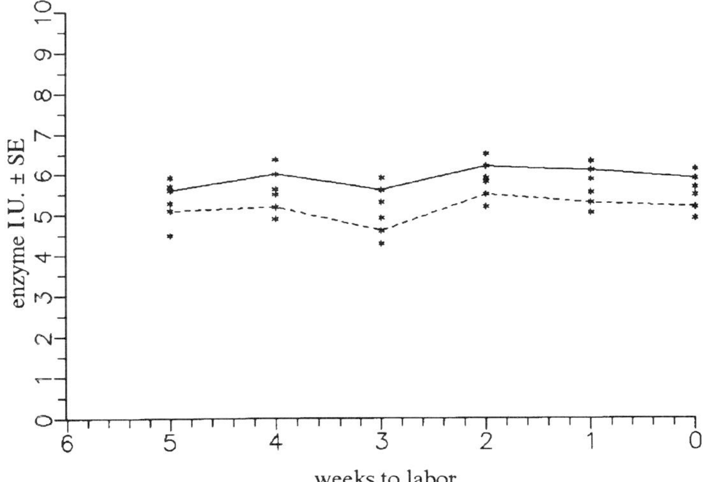

Figure 5.4 Mean values and their standard error of irregular profile of oxytocinasemia (—) and isooxytocinasemia (- -) in pregnant women with hypertension

Table 5.1 Oxytocinasemia ($\bar{x} \pm$ SD) for biological gestational age in pregnant women with diabetes mellitus (μmol/l/min)

| Group | Weeks to labor | | | | | |
	0	1	2	3	4	5
A	8.0 ± 2.7 (50)	7.4 ± 2.8 (41)	6.6 ± 2.2 (25)	5.9 ± 1.7 (17)	5.0 ± 0.9 (6)	5.0 ± 1.0 (13)
B	7.3 ± 2.7 (48)	7.5 ± 2.1 (50)	7.1 ± 2.0 (42)	6.7 ± 2.8 (25)	6.7 ± 3.4 (23)	6.2 ± 1.4 (9)
C	6.8 ± 2.0 (100)	6.8 ± 2.3 (108)	6.0 ± 2.0 (85)	5.7 ± 1.8 (64)	4.9 ± 2.0 (52)	4.9 ± 1.5 (41)
A + B + C	7.2 ± 2.4 (198)	7.1 ± 2.4 (199)	6.4 ± 2.1 (152)	6.0 ± 2.1 (106)	5.4 ± 2.5 (81)	5.1 ± 1.5 (63)

(N): number of measurements.

Table 5.2 Isooxytocinasemia ($\bar{x} \pm$ SD) for biological gestational age in pregnant women with diabetes mellitus (μmol/l/min)

| Group | Weeks to labor | | | | | |
	0	1	2	3	4	5
A	6.5 ± 2.5 (27)	5.9 ± 2.6 (23)	5.9 ± 1.8 (10)	4.4 ± 0.3 (4)	4.6 ± 0.0 (2)	3.7 ± 0.1 (3)
B	6.4 ± 2.6 (31)	5.9 ± 1.8 (26)	5.6 ± 1.4 (28)	5.4 ± 2.4 (16)	5.6 ± 3.7 (13)	3.7 ± 0.4 (2)
C	5.8 ± 1.6 (56)	5.6 ± 1.7 (57)	5.1 ± 1.4 (46)	4.8 ± 1.4 (40)	4.9 ± 1.5 (34)	4.5 ± 1.3 (26)
A + B + C	6.1 ± 2.2 (114)	5.7 ± 1.4 (106)	5.2 ± 1.5 (84)	5.0 ± 1.7 (60)	4.7 ± 2.2 (49)	4.4 ± 2.2 (31)

(N): number of measurements.

throughout two last gestational weeks; C, with an irregular, variable gestational profile of the enzyme level (Figures 5.2–5.4).

Tables 5.1 and 5.2 show the results of oxytocinase and isooxytocinase determination in diabetic patients during consecutive weeks before labor. The last 6 days before labor form period 0. The prepartal oxytocinase level 7.2 ± 2.4 IU was significantly lower than that observed in healthy women (8.2 ± 2.2, $t = 3.8$, $p < 0.001$), and the average values obtained during the preceding weeks were also lower. While comparing the isooxytocinase levels such differences were not observed, As expected the levels of both isoenzymes were lowest in sub-group C.

Tables 5.3 and 5.4 present the results obtained in women with hypertension. These results, though slightly lower in enzyme level, appeared to be similar to those seen in women with diabetes. However, the oxytocinase level ceased to increase 2 weeks before labor unlike in diabetic and especially healthy women, similarly the level of isooxytocinase reached a plateau 2 weeks earlier.

The enzymatic profiles in both comparative groups are different. In patients with vascular changes due to diabetes the linear correlation between oxytocinase and isooxytocinase level and newborns' body weight (x) was stated (respectively: $y = 0.18x + 1.8$, $r = 0.62$, $t = 4.7$, $p < 0.001$ and $y = 0.33x + 1.2$, $r = 0.66$, $t = 3.9$, $p < 0.001$), whereas in patients with hypertension the linear correlation only between oxytocinase level and newborns' body weight $(r = 0.32, t = 8, p < 0.01)$ was observed. These relations stress the different character of etiopathogenesis of the observed gestoses.

It should be emphasized that in healthy women the continuous increment of blood oxytocinase level until labor is observed in 81% of cases (Figure 5.2), it levels off or decreases several days before labor in 12% (Figure 5.3), and is irregular, with variable profile in 7% [6]. In patients suffering from diabetes or hypertension the percentages tend to behave in an opposite way and they equal 27.1%, 27.1%, 45.8% and 16%, 24%, 60%, respectively (Figure 5.4).

In both study groups the degree of the decrease and deflection of the oxytocinase profile, as compared with the data observed in healthy women, constitutes a measurement of the development of neurohormonal perturbations. This remains coherent with the statement that in pregnant women threatened with miscarriages or premature deliveries, the values of oxytocinase decreased or were stationary. However ACTH therapy in such cases makes a successful outcome of pregnancy possible[23,82–88].

NEUROHORMONAL GESTOSIS

Hypothalamic hormones induce the production of placental and tissue oxytocinases. Therefore, pregnancy monitoring using these two enzymes is especially significant in the case of women with neuroendocrinological gestosis, i.e. women treated because of infertility in the past, treated because of juvenile or post-preg-

Table 5.3 Oxytocinasemia ($\bar{x} \pm$ SD) for biological gestational age in pregnant women with hypertension (μmol/l/min)

Group	Weeks to labor					
	0	1	2	3	4	5
A	8.6 ± 2.0 (11)	7.6 ± 1.5 (14)	6.9 ± 1.4 (7)	7.8 ± 1.4 (5)	6.3 ± 1.3 (2)	5.6 ± 0.9 (4)
B	6.5 ± 2.1 (23)	7.0 ± 2.5 (23)	7.1 ± 2.0 (16)	7.0 ± 2.4 (12)	5.8 ± 1.4 (7)	(0)
C	5.9 ± 1.8 (59)	6.1 ± 1.9 (67)	6.2 ± 2.0 (44)	5.6 ± 1.8 (34)	6.0 ± 1.5 (16)	5.6 ± 1.2 (14)
A + B + C	6.3 ± 2.1 (93)	6.5 ± 2.1 (104)	6.5 ± 2.0 (67)	6.1 ± 2.0 (51)	6.0 ± 1.4 (25)	5.6 ± 1.1 (18)

(N): number of measurements.

Table 5.4 Isooxytocinasemia ($\bar{x} \pm$ SD) for biological gestational age in pregnant women with hypertension (μmol/l/min)

Group	Weeks to labor					
	0	1	2	3	4	5
A	6.1 ± 0.8 (3)	5.9 ± 0.6 (5)	5.6 ± 0.0 (1)	5.6 ± 0.3 (2)	(0)	(0)
B	5.3 ± 1.0 (3)	5.2 ± 0.4 (3)	4.9 ± 0.2 (3)	5.1 ± 0.3 (3)	(0)	(0)
C	5.2 ± 1.6 (31)	5.3 ± 1.5 (36)	5.5 ± 1.5 (23)	4.6 ± 1.4 (19)	5.2 ± 1.0 (11)	5.1 ± 1.2 (4)
A + B + C	5.2 ± 1.5 (37)	5.3 ± 1.4 (44)	5.4 ± 1.4 (27)	4.8 ± 1.3 (24)	5.2 ± 1.0 (11)	5.1 ± 1.2 (4)

(N): number of measurements.

nancy hypothalamic syndromes or those who have taken contraceptive pills for prolonged periods of time.

The ACTH replacement therapy, introduced by R. Klimek[2,83] in the 1960s led to a decrease in the number of intrauterine fetal deaths from 80% to <10% in women with hypothalamic syndrome.

The clinical application of ACTH as an optional treatment in hypothalamic insufficiency syndrome has proved successful in preventing premature labors, neonatal respiratory distress

Table 5.5 Outcome of pregnancy in patients with ACTH therapy ($\bar{x} \pm$ SD)

Variables N	140	Variables N	140 (100%)
Mother		Labor	
age (years)	29.7 ± 5.3	spontaneous	116 (83%)
gravidity	3.1 ± 1.3	induced	24 (17%)
parity	2.2 ± 1.1	natural	104 (74%)
abortion	1.0 ± 1.1		
children	0.7 ± 0.8		
gestation (wks)	37.5 ± 4.6		
Newborn		McDonald suture	43 (31%)
weight (kg)	2.88 ± 0.91	Tocolysis	61 (43%)
length (cm)	50.6 ± 7.10	Hemorrhage	26 (18%)
Apgar scores	9.0 ± 2.00		
stillborn	10 (7%)		

Table 5.6 Oxytocinase and isooxytocinase ($\bar{x} \pm$ SD μmol/l/min) for calendar gestational age in women treated with synthetic ACTH

Weeks postmenstrual	Oxytocinase	t	Isooxytocinase	t
>40	6.5 ± 2.2 (19)	NS		—
40–37	6.6 ± 2.0 (147)	*** 7.7	5.0 ± 1.6 (23)	** 2.7
36–33	5.0 ± 1.8 (191)	*** 6.8	4.0 ± 1.4 (58)	*** 3.4
32–29	3.9 ± 1.5 (225)	*** 10.3	3.3 ± 1.1 (92)	*** 4.5
28–25	2.7 ± 0.8 (213)	*** 7.3	2.6 ± 0.7 (64)	NS
24–21	2.1 ± 0.6 (129)		2.3 ± 0.6 (22)	

(N): number of measurements; NS: not significant; * $p < 0.05$, ** $p < 0.01$, *** $p < 0.001$.

Table 5.7 Oxytocinase and isooxytocinase ($\bar{x} \pm$ SD µmol/l/min) for biological gestational age in women treated with synthetic ACTH

Weeks to labor	Oxytocinase	t	Isooxytocinase	t
0	5.5 ± 2.1 (48)	NS	4.0 ± 1.0 (8)	NS
1–2	5.6 ± 2.3 (181)	*** 3.5	4.1 ± 1.5 (59)	NS
3–4	4.7 ± 1.8 (107)	* 2.2	3.6 ± 1.3 (43)	* 2.3
5–8	4.2 ± 2.0 (199)	*** 3.9	3.0 ± 1.2 (54)	NS
9–12	3.5 ± 1.5 (191)	*** 7.2	3.4 ± 1.3 (53)	*** 3.6
13–16	2.5 ± 0.8 (142)	*** 5.2	2.5 ± 0.7 (32)	NS
17–20	2.0 ± 0.5 (85)		2.3 ± 0.5 (20)	

(N): number of measurements; NS: not significant; * $p < 0.05$, ** $p < 0.01$, *** $p < 0.001$.

syndrome, as well as pernicious vomiting etc.[89–96]. After a decade it has been supported by worldwide corticotherapy used as the substitute[97–114].

M. Szlachcic[115] compared the results of oxytocinase and isooxytocinase determinations obtained from a group of 140 pregnant women with an average age of 29.7 ± 5.3 years (range 19–42 years). The patients had undergone treatment with a synthetic ACTH of prolonged activity (Cortrosyn-Depot Organon, Synacthen-Depot Ciba-Geigy), administered three times in 0.5 mg doses every other day. The mean parity of the mothers was 3.1 ± 1.3, ranging from 1 to 7. These women had experienced on average 2.2 ± 1.1 (1–5) labors and 1.0 ± 0.1 (0–5) miscarriages and had until the current pregnancy only 0.7 ± 0.8 (0–4) children; this number constitutes a 30% successful outcome of previous pregnancies (Table 5.5) reflecting the poor prognosis of pregnancy in this group if untreated.

The pregnancies studied lasted for 37.5 ± 4.6 weeks (range

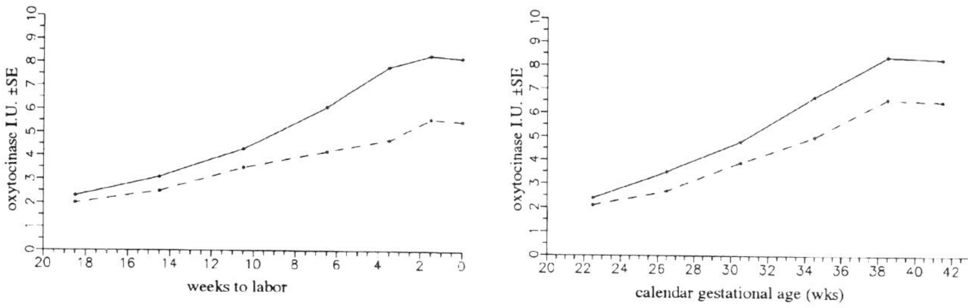

Figure 5.5 Comparison of gestational oxytocinase profiles of healthy (—) and treated (- -) pregnant women

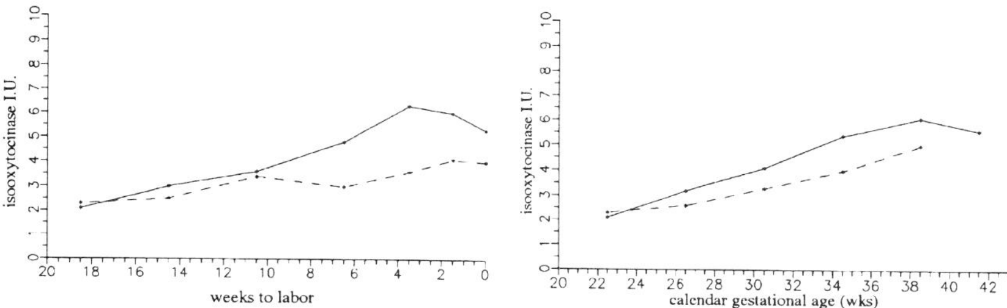

Figure 5.6 Comparison of gestational isooxytocinase profiles of healthy (—) and treated (- -) pregnant women

21–42), resulted in the birth of 130 healthy infants, with average Apgar score 9.2 ± 2.0 (5–10), body weight 2880 ± 910 (1100–4370) g and body length 50.6 ± 7.1 (41–60) cm; 10 cases of stillbirth were observed (prenatal mortality, 3; perinatal mortality, 7). In 116 cases (83%) spontaneous labors were observed, and the number of premature labors was 24 (17%).

Tables 5.6 and 5.7 show the oxytocinase and isooxytocinase values (IU = μmol/l/min) in particular weeks of the calendar and biological time of pregnancy duration.

In accordance with previous papers[65,67,71,116], it was again confirmed that oxytocinase level against calendar age shows a statistically significant increment in weekly intervals until the second half of biological range of labor occurrence and remains stable at the level achieved in the first half of biological range of labor occurrence (6.6 ± 2.0 and 6.5 ± 2.2 IU). If we take into account the biological state of pregnancy we can also say that in weekly intervals the oxytocinase level increment is statistically significant and reaches the highest values several days before labor (5.5 ± 2.1 and 5.6 ± 2.3 IU). Also the increment of isooxy-

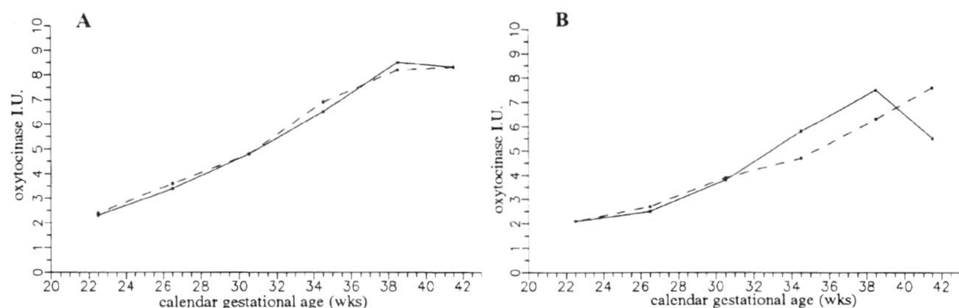

Figure 5.7 Comparison of calendar gestational oxytocinase profiles of healthy (A) and treated (B) pregnant women in relation to parity: primipara (—) and multipara (- -)

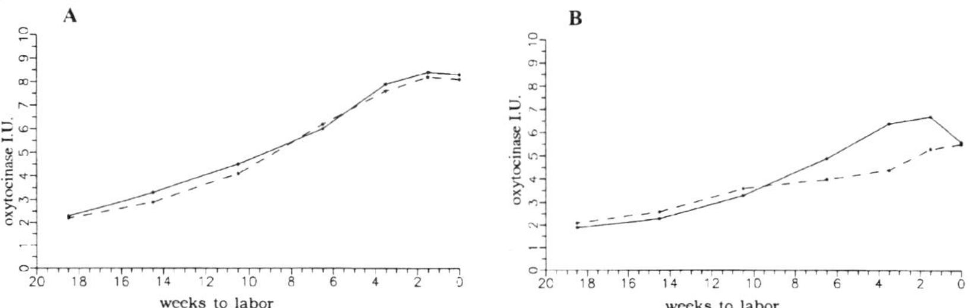

Figure 5.8 Comparison of biological gestational oxytocinase profiles of healthy (A) and treated (B) pregnant women in relation to parity: primipara (—) and multipara (- -)

tocinase level until the biological range of human labor occurrence (5.0 ± 1.6 IU) was confirmed. The fact that isooxytocinase level reaches the highest plateau as early as one month before labor (3.6 ± 1.3 IU) was proved.

If we compare women requiring adrenocorticotherapy and healthy women we see that levels of both isoenzymes are lower in the group requiring therapy in aspect of calendar pregnancy duration as well as biological condition, and, what is more, isooxytocinase levels reach a plateau several months before labor. The above results are presented in Figures 5.5 and 5.6 where levels of both isoenzymes are compared with those in healthy women[34,68,81].

The group under investigation consisted of 89 multiparae and 51 primigravidae. They did not differ in newborn condition, pregnancy duration time, number of labor inductions and its outcome or the necessity of oxytocin or tocolitics usage. It seemed interesting to examine the relationship between parity and oxytocinase and isooxytocinase profiles. Statistically signi-

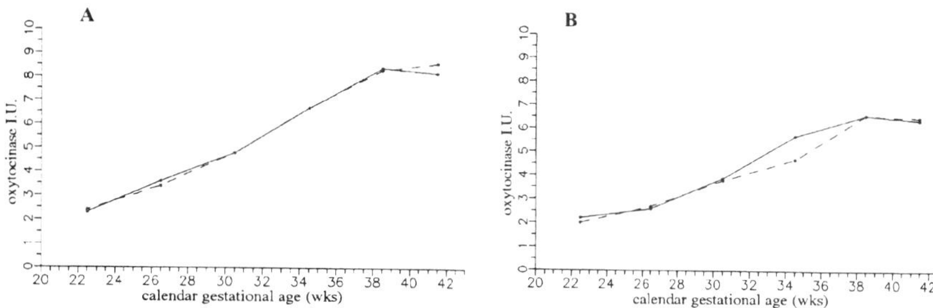

Figure 5.9 Comparison of calendar gestational oxytocinase profiles of healthy (A) and treated (B) pregnant women in relation to mother age: ≤ 27 (—) and >27 (- -) years

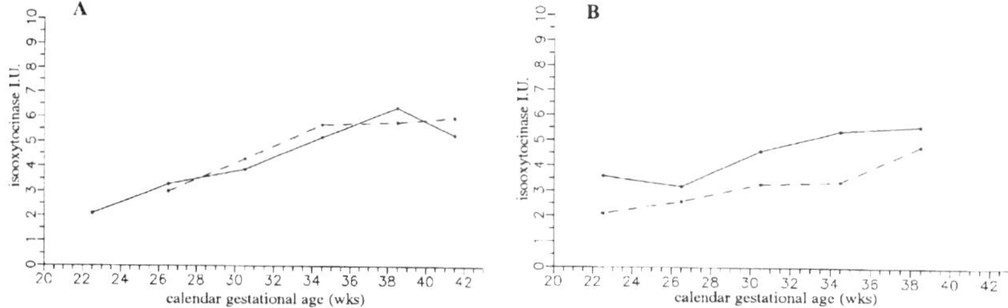

Figure 5.10 Comparison of calendar gestational isooxytocinase profiles of healthy (A) and treated (B) pregnant women in relation to mother age: ≤ 27 (—) and >27 (- -) years

ficant increments in oxytocinase level until the end of observation in weekly intervals was shown both in multiparae and primigravidae. In weeks of pregnancy duration the oxytocinase level in primigravidae is statistically significantly higher from 33 to 40 calendar week of pregnancy duration in comparison with multiparae ($t = 3.9$ and $t = 3.3$, $p < 0.001$), but the opposite occurs in the second half of the biological range of labor occurrence. It appears that oxytocinase level in primigravidae treated with ACTH is statistically significantly higher than in multiparae by 2 months before labor ($t = 2.9$, $p < 0.001$) and as late as in the last few days these differences are equalized. This concerns multiparae maintaining the oxytocinase level and in primigravidae the decrease of oxytocinase level during the last few days before labor (Figure 5.7b). This phenomenon was not observed by A. Michalski in healthy patients[67]. The significance of those differences is the more important since we deal here with patients with *a priori* low level of enzymes, requiring adrenocorticotherapy (Figures 5.7 and 5.8).

The analysis of the data according to the age of the women was also carried out. Again patients were divided into groups according to their age (Figures 5.9 and 5.10). It was shown that levels of tissue oxytocinase are much more variable and independent than placental oxytocinase. In women below 27 years of age the enzyme values remain at a higher level than in older women although these two groups do not differ in onset of labor, its course or necessity of therapy during the whole pregnancy.

In conclusion it should be stressed that oxytocinase and isooxytocinase levels in pregnancies where ACTH therapy was applied are statistically significantly lower than in healthy women. The enzyme profiles are also different. Blood levels of oxytocinase and isooxytocinase often are constant or decreasing in women requiring any therapeutic procedure.

When determining oxytocinase in pregnant women we should also take into account not only the obstetric history of patients with hypothalamic syndromes but also their age and gravidity. In healthy women neither the number of children nor their age influences oxytocinase levels. In women with hypothalamic syndromes, however, such differences are statistically significant. This gives a new insight into the problems of threatened pregnancy monitoring and more important, shows that oxytocinase levels in women treated with ACTH are statistically significantly lower than those obtained in women suffering from diabetes or hypertension.

6
Biochemical methods of oxytocinase determination

Alicja Rzepecka and Rudolf Klimek

LABORATORY PITFALLS OF OXYTOCINASE DETERMINATION

As far back as 1930 it was demonstrated by the Hungarian physician, K. Fekete (*Endokrinologie*, 1930;**7**:354) that the serum of pregnant women was capable of abolishing the uterus-stimulating activity of posterior pituitary extracts. It was not till the late forties that this feature of the blood of pregnant women was linked with enzyme activity. The enzyme was called oxytocinase (E.W. Page, E. Werle, G. Effkeman, A. Hevelke, K. Buthmann). At the beginning of enzyme research, oxytocin or vasopressin activities were measured before and after incubation with biological fluids or tissue extracts containing oxytocinase. The results obtained were divergent and corresponded poorly with the clinical findings, so the biological methods were not used in clinical practice.

The next step was made in 1957 by H. Tuppy and H. Nesvadba[117] who described the principles of chemical assay of the enzyme. They used a synthetic substrate L-cystine-di-β-naphthylamide. Unfortunately, the authors made a mistake in their description of the method, indicating erroneously a 1% concentration of sodium nitrate solution instead of 0.1% which precluded the determination of oxytocinase. The procedure was repeated in 1961 by R. Klimek and M. Pietrzycka, who revised the methodological error and made the first attempt at establishing an international unit of enzyme activity: μmol/l/min. (*Clin. Chim. Acta*, 1961;**6**:326).

The following years brought an increasing interest in biochemical and clinical assessment of aminopeptidase activity in obstetrics (B. Berde, J. Berhard, K. Drewniak, P. Fylling, T. Hashimoto, R. Klimek, I. Holanska-Nalepa, S.E.H. Melander, Z.B. Miller, W. Muller-Hartburg, H. Nesvadba, A.P.M. van Oudheusden, A.M. Riad, I. Rychlik, G. Ryden, A. Rzepecka, K. Semm, I. Sjoholm, J. Stanek, F. Sorm, M.A. Titus, H. Tuppy *et al.*). However, some of these researchers made serious mistakes both in the chemical process of enzyme estimation and in the clinical interpretation of results. For example, in 1966 C. Babuna and E. Yenen published a study on the modification of the original chemical assessment of oxytocinase (*Am. J. Obstet. Gynecol.*, 1966;**94**:868 and **95**:925). Substantial methodological errors and misunderstanding of the results led the authors to wrong conclusions, which in fact hindered investigations of enzyme monitoring in pregnancy. Nevertheless three years later R. Klimek and M. Malolepszy[78] delivered a paper in which they corrected these laboratory mistakes. Unfortunately it was published in a biochemical journal which was not readily available to clinicians.

Moreover, the use of another substrate L-leucyl-β-naphthylamide for determination of oxytocinase activity in pregnant women led to the assessment of total aminopeptidase activity instead of oxytocinase alone. It is well known that some pathological conditions, e.g. impaired hepatic function, may influence aminopeptidase activity. Thus, this kind of enzyme estimation may not properly reflect the placental function and interferes with clinical implications[10,65,71,90,118–124].

Finally, the introduction of new methods of monitoring the course of pregnancy, e.g. human placental lactogen, ultrasonography as well as divergences of opinion on oxytocinase application resulted in neglect of the study of cystine-aminopeptidase activity in the serum during pregnancy.

However, the continuation of our investigation on the use of various substrates and buffers resulted in clinical verification of chemical methods that could be applied in the monitoring of pregnancy and for the prediction of delivery.

OXYTOCINASE DETERMINATION

Knowledge of the structure and the synthesis of oxytocin and its analogs made it possible to elucidate the mechanism of oxytocin inactivation by oxytocinase. H. Tuppy and H. Nesvadba[117] found serum oxytocinase to be an aminopeptidase, by which the cyst(e)inyltyrosine bond is broken. This reaction has been confirmed by F. Sorm and co-workers[178] who used synthetic substrates (oxytocin components or their derivatives). Tissue oxytocinase was also shown to act in the same way[33,59,64,90].

These studies made possible the synthesis of substrates with which oxytocinase determination could be determined without application of oxytocin. In addition to L-cystine-di-β-naphthylamide, S-benzylcysteinyltyrosine and S-benzylcysteinyl-glycineamide, more recently L-cysteine-di-*p*-nitroanilide and *S*-benzyl-L-cysteine-β-naphthylamide have been used. By a comparison of these methods with biological ones, the identical activities of aminopeptidase and oxytocinase were shown both in non-purified serum oxytocinase, and in highly active serum preparations obtained by electrophoresis, ionites, salting out, or alcohol fractionation. Oxytocinase A (in albumin fraction) and B (in globulin fraction), isolated by Czech authors from the serum of pregnant women, are probably two forms of the same enzyme. In spite of physical and chemical differences, they are similar in their specificity to the non-purified serum oxytocinase but differ significantly from tissue oxytocinase and leucyl-amino-peptidase.

Finally, E.W. Page and co-workers[61], using agar gel electrophoresis, revealed two cysteine-aminopeptidases in the serum, present only in pregnancy. After concentrating 4500 times, the oxytocinase from the serum of extraplacental blood decomposed many β-naphthylamide aminoacid derivatives even more actively than L-cystine-di-β-naphthylamide. It is therefore not a specific cystineaminopeptidase. Peptides and amides containing cysteine with an amino endgroup are not excluded from its action. In this respect it differs significantly from other aminopeptidases, e.g. leucine-aminopeptidases (LAP). The increase in LAP activity, observed by other authors, is due to the ability of the oxytocinases being formed to decompose leucylamide, but not vice versa.

The use of chemical methods for oxytocinase determination affords a means of observing the effects of inducing specific enzymes by increasing endogenous hormone or its analog. Conversely, direct action on oxytocinase allows observation of the effects of hormones under new conditions. Both procedures allow insight to be gained into complex reproduction mechanisms, thus further improving our practical measures for assisting labor.

Laboratory and clinical repeatability of any clinical method constitutes the essential criterion of its usefulness[59,61,63,64,70,90,116,117,125–141]. The sensitivity, specificity, accuracy, and prognostic value of each method, especially chemical, can be strictly determined. However, it is the clinical condition of a patient that most interests the physician. The doctor himself should have an idea of the probable results of the test from his clinical evaluation before sending the sample to the laboratory. He should assume beforehand that the results of biochemical determinations in pregnant patients with hypertension, hypothalamic syndrome or diabetes will be lower than in healthy pregnant women. The stability of this phenomenon proves to be of extremely great importance.

The oxytocin–oxytocinase system (0–0.S) is a stable regulatory and informational mechanism in a human body. This can be proved by evaluation of oxytocinase and isooxytocinase every other day. But in clinical practice it is difficult to collect a sufficient amount of biochemical data since the enzymatic evaluation is justified only in cases with very well defined clinical indications only, i.e. cases in which such evaluation is necessary from the clinical point of view and when it directly serves the patient. Nevertheless many years of documentation have enabled us to collect a register of 517 cases in which the enzymatic levels were determined three times a week in the second and third trimester of pregnancy[66].

In 269 cases (52%), three successive results showed no more variability than 0.2 IU, which proves the stability of the observed mechanism and the chosen method. In the next 154 cases (30%), an increase of more than 0.2 IU was observed although it did not exceed 0.5 IU, whereas in the next 94 cases (18%), an analogical decrease of the enzyme level was noticed. The investigation of periods of pregnancy in which the increase or decrease of blood enzyme activity was observed, seems most interesting. It appeared that the determination was performed within the period of 2 weeks before labor in the first case in 36% and in the second in 41%. In pregnant women with stable enzyme levels the amount of the above determination was half as much. What is more, in the group characterized by a decrease of blood oxytocinase level, the number of preterm labors was almost double (23%) in comparison with cases with a stable level (14%) or an increasing one (14%).

The above observations were based on non-selective data; the study also included pregnant women with an irregular enzymatic profile for the entire duration of pregnancy. Fortunately pregnancy is a natural physiological condition. This is why a doctor calling upon knowledge of the clinical condition of his patients can and should be able to assess the methodological accuracy of biochemical determinations. If the numerous laboratory results tend to reveal one-directional diversion, proportional to their absolute values, it is usually due to a change in pH of the enzyme solution, caused by lengthening or shortening of the incubation time even only by several minutes. Therefore a doctor should be aware of the principles relevant to the biochemical methods.

The method of H. Tuppy and H. Nesvadba, modified by R. Klimek[2] has been constantly applied in our department for over 30 years (presented below). Dozens of determinations of the two enzymes are performed three times a week (Mondays, Wednesdays, and Fridays). The determination is performed in this way to allow enough time for the final reaction and the concentration of its products to be deciphered the next morning. If immediate results are needed the incubation time may be halved and the results may be deciphered after 2 hours. It is also possible to take blood samples with an interval of 2–4 h and determine the possible direction of change of the enzymic activity.

Biophysical examinations record the state of the fetus at a certain time only and do not exclude the sudden future appearance of a pathological syndrome. This considerably limits their prognostic usefulness. This is not the case when using enzymatic determinations. Separate results reveal the level of endogenous production of cyclopeptide hormones. The concentration changes not only provide information about pregnancy development, fetal and placental maturation, but also allow the prediction of future placental insufficiency and hence possible fetal distress.

CHEMICAL DETERMINATION OF SERUM OXYTOCINASE AND ISOOXYTOCINASE

Reagents

(a) Substrate: dissolve 135 mg L-cystine-di-β-naphthylamide in 50 ml 0.012 N HCl, with moderate warming and filtrate, then add H_2O to volume of 100 ml, store in the refrigerator.

(b) Barbitone buffer (pH 7.9) for oxytocinase determination: dissolve 2.07 g sodium barbitone in 100 ml bidistilled H_2O (free of CO_2) to make 0.1 N solution. To 68.9 ml of such solution add 31.1 ml 0.1 N HCl and 50 ml bidistilled H_2O giving a buffer of pH = 7.9;

(c) Phosphate buffer (pH 6.2) for isooxytocinase determination: solution A: dissolve 3.6312 g KH_2PO_4 in 400 ml bidistilled H_2O; solution B: dissolve 2.386 g $Na_2HPO_4.12H_2O$ in 100 ml H_2O. Mix 400 ml of solution A with 100 ml of solution B to give a buffer of pH 6.2.

(d) 10% trichloroacetic acid;

(e) 0.1% $NaNO_2$

(f) 0.5% ammonium sulfamate;

(g)   0.1% *N*-(1-naphthyl)-ethylendiamine-dihydrochloride
(h)   Acetone–acid mixture (two volume 0.36 N HCl and one volume reagent grade acetone);

Procedure

Into test-tubes for centrifuge add in turn: 0.15 ml H_2O, 0.5 ml buffer, 0.1 ml serum (without hemolysis!), then mix well. Add 0.25 ml substrate mixing solution again. Place test-tubes at 37°C for 2 h and then stop reaction with 1 ml 10% trichloroacetic acid. Shake, centrifuge and decant. Add 1 ml of each decanted solution to 5 ml acetone–acid mixture placed in 100 ml conical flasks.

The next steps are performed in a darkroom. Mixing solutions in conical flasks add in 3 min intervals: 1 ml 0.1% $NaNO_2$, 1 ml ammonium sulfamate and 1 ml ethylendiamine. Then flasks are left at 37°C for 3 h. The maximal color is reached after 3 h and remains constant for several hours. Spectrophotometrical measurements are performed at 565 nm for control solution which is composed of 5 ml acetone-acid mixture, 1 ml $NaNO_2$, 1 ml ammonium sulfamate and 1 ml ethylenediamine. It is not necessary to get control samples for each individual determination unless the patient is treated with sulfonamides or nitrofurantoins.

The standard curve is made with standard solution: 40 mg β-naphthylamine dissolved in 500 ml 0.012 N HCl while warming and stirring, then dilute to one liter with bidistilled water. Store in the dark.

* * *

It is necessary to determine both isoenzymes (CAP_1 and CAP_2) because in addition to clinically useful information it also provides a control of the methodology. Incubational solutions with an optimum of pH 7.9 and 6.2 respectively, have been chosen and any diversion from these values leads to a decrease of CAP_1 and an increase of CAP_2 activity.

The mean ±SD for serum oxytocinase values after 14 days measurement at: room temperature, 4°C and –20°C are not statistically different and their corresponding coefficients of variation are 5.2, 5.0 and 5.0%, respectively. Also the short-term fluctuations of serum oxytocinase levels over 150 min are not statistically different, with mean coefficients of variation of 3.6%, i.e. a variability within the 5% error of the method[130].

7

Enzymatic and ultrasonographic prediction of birth-date

Marek Klimek and Rudolf Klimek

Reports about monitoring the development of pregnancy and predicting the delivery term abound in the literature and the techniques have become firmly established in obstetrics[15,18,21,25,56,59,125,142–168]. Unfortunately, the possibilities they offer have been considerably limited because interpretation is based on the calendar scale of time. If the delivery of a mature baby occurs physiologically within the calendar norm of 6 weeks, two fetuses of the identical mass and length, can have the same biological age but their calendar age, found on the daily, weekly or monthly calendar scale, may range from 37⁰/7 up to 43²/7 weeks (Figure 7.1). In this example the body weight of the unborn baby, who matures by the 259th day of calendar pregnancy, is one-third higher than that of an infant, who being at the same calendar age will reach biological maturity

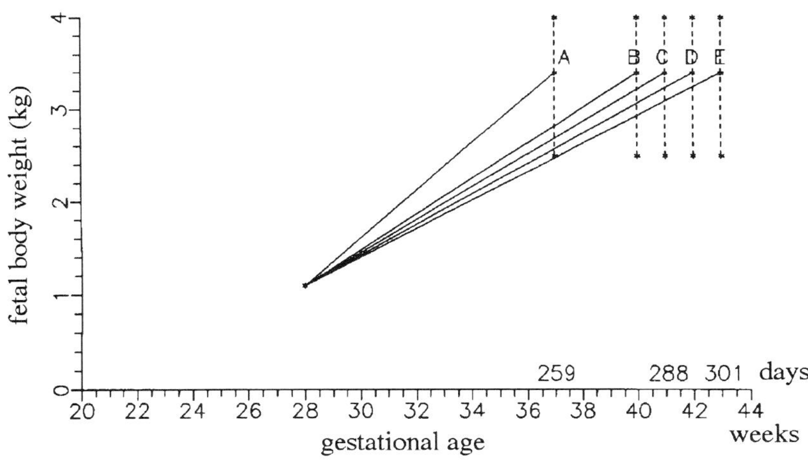

Figure 7.1 Increase of fetal body weight for calendar gestational age

[53]

for delivery within the next 6 weeks, i.e. on the 303rd day.

Unfortunately, according to ACOG Committee Opinion 98[36] fetal maturity has been assumed, if one of the following criteria is met:

(1) Fetal heart tones have been documented for 20 weeks by non-electronic fetoscope or for 30 weeks by Doppler;
(2) It has been 36 weeks since a positive pregnancy test was performed by a reliable laboratory;
(3) An ultrasound measurement of the crown–rump length, obtained at 6–11 weeks, supports a gestational age 39 weeks, or
(4) An ultrasound, obtained at 12–20 weeks, confirms the gestational age of 39 weeks determined by clinical history and physical examination.

Therefore, according to the above opinion it is thought appropriate to carry out an elective Cesarean delivery at 39 weeks by menstrual dates. In its defence the document correctly, but not only from the medical point of view stated, 'that awaiting the onset of spontaneous labor is another option'.

Oxytocinase determination can be used to improve our understanding, illustrating the proper interpretation of the obstetric data. The average value of oxytocinase for mature fetuses, normally born after spontaneous onset of labor is

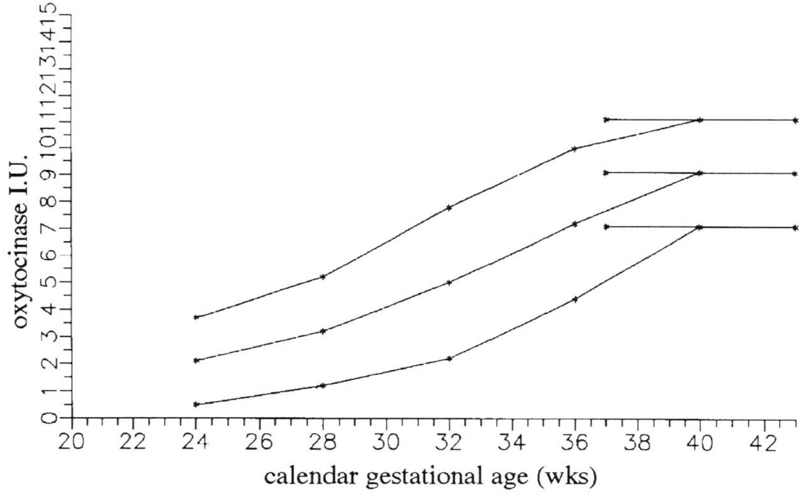

Figure 7.2 Mean values of oxytocinasemia and their standard deviation throughout advanced gestation

$9.1 \pm 2.0\ \mu\text{mol}/1/\text{min}$ from $37^{0/7}$ to $43^{2/7}$ weeks of the calendar scale of pregnancy development (Figure 7.2).

Figure 7.2 represents the calendar scale from $20^{0/7}$ to $43^{2/7}$ weeks of pregnancy and the average values of oxytocinase and two standard deviations which signify the biological state of the fetus or pregnancy. Week intervals are marked on the axis of abscissae, the axis of ordinates includes the analyzed values.

The average values $\pm 2\text{SD}$ for the $24^{0/7}$ week indicate the ranges of correct values that during the subsequent weeks become connected with the enzyme values concerning mature fetuses. The diagram shows that beginning from the 28th week, the average oxytocinase increase is linear in character ($y_x = 0.5x - 10.8$). The lines of the two standard deviation values ($y_{-2D} = 0.5x - 13.2$ and $y_{+2D} = 0.5x + 8.4$, correspondingly) mark the norm's range.

Not only the line that links the average value of the 28th week with the middle of the calendar norm (regular fetal growth), but also the lines connecting it with the beginning (fast fetal growth) and end (slow fetal growth) of the normal births range are equally important (Figure 7.3).

In order to facilitate the reading of the calendar scale (20–43 weeks), for delivery-date prediction, a scale for the pregnancy biological age (15–0 weeks before labor) has been added (Figures

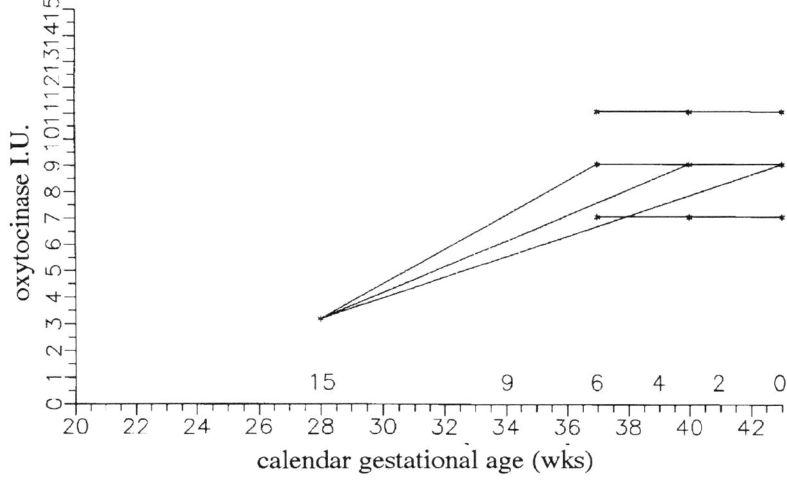

Figure 7.3 Average increase of oxytocinasemia for calendar (20–43 weeks) gestational age

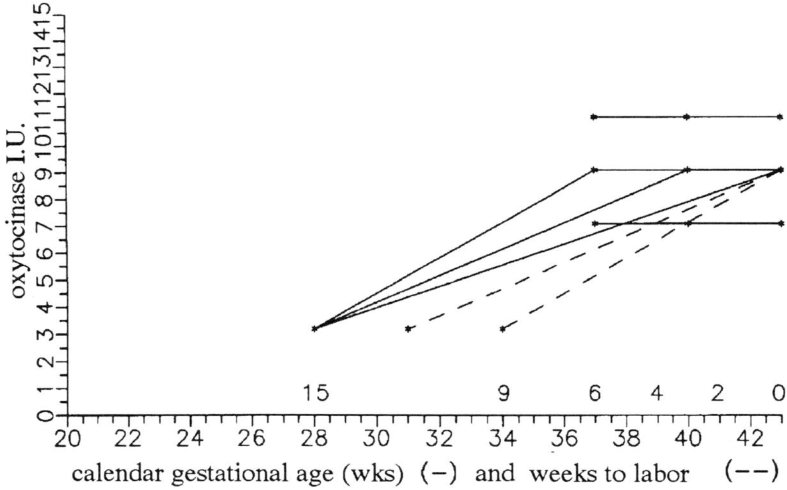

Figure 7.4 Mean oxytocinasemia increase for biological (15 – 0) and calendar (20 – 43) gestational age

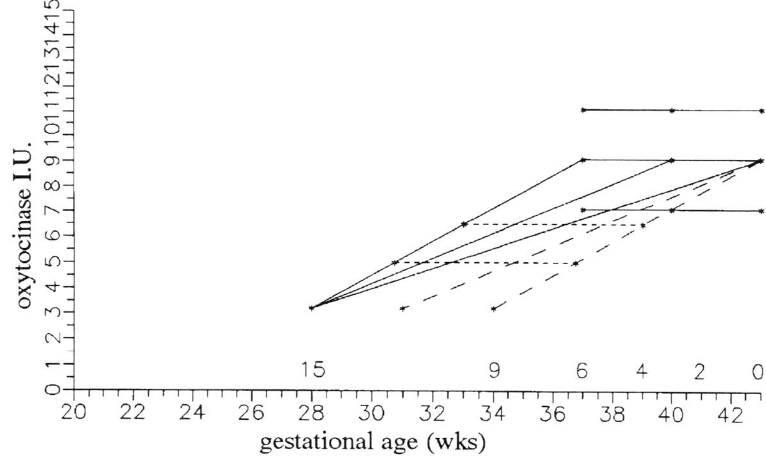

Figure 7.5 Prediction of confinement using biological–calendar scale for oxytocinase measurements

7.3 and 7.4). The average value for the birth at the end of the normal range, has been linked with lines that determine the increase of the analyzed value at the beginning, in the middle and at the end of the calendar norm, separately. Their three regression equations are as follows: $y = 0.4x - 6.6$; $y = 0.5x - 10.6$; $y = 0.6x - 17.1$. The shifting of the end of the biological scale

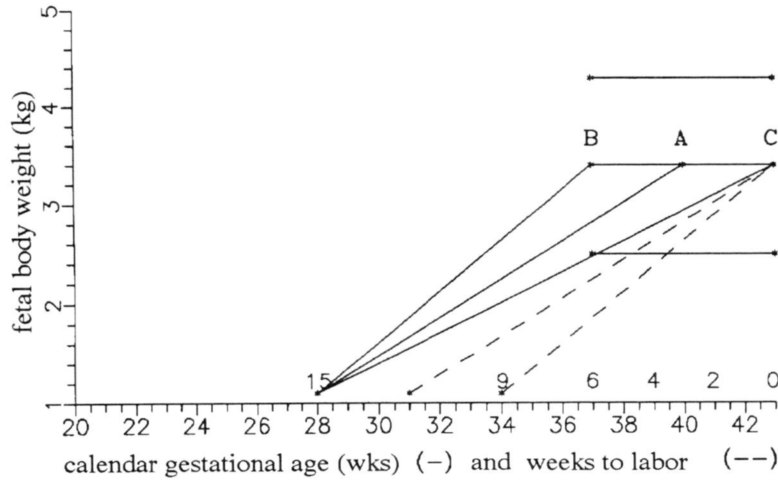

Figure 7.6 Fetal body weight gain for biological–calendar scale

(0) to the end of the calendar scale (43rd week) enables the evaluation of the baby's maturity for labor, as well as the pregnancy duration from the last menstrual period.

Making use of both scales (Figure 7.5) a chosen oxytocinase value, e.g. 5.0 IU is both equivalent either to a range of the 30⁵/₇–32³/₇ calendar weeks or to a biological age of 10 to 6 weeks before labor. After two weeks the subsequent measurement, e.g. 6.5 IU depicts the progress of pregnancy with new ranges of 33⁰/₇–36²/₇ calendar weeks and 7–4 weeks to delivery. It is now possible to predict the labor term more accurately by comparing these two ranges and by taking into account the period between the enzyme determination. In this example the probable labor will take place after 4 weeks (line of the quickest increase of enzyme level).

The high correlation between oxytocinase level and fetal weight is commonly known[2,13,59,74–76,85,122,159,169,170]. Therefore, a diagram identical to Figure 7.4 has been used by M. Klimek[6] to show the classical linear increase of body weight and length after 28 weeks of pregnancy duration (Figure 7.6).

Figures 7.7–7.10 show the head and abdominal circumferences, biparietal diameter and femur length, on the calendar–biological scale for the proper interpretation of ultrasound results.

This joint biological–calendar scale, introduced to evaluate pregnancy development, forms the link between the enzymatic

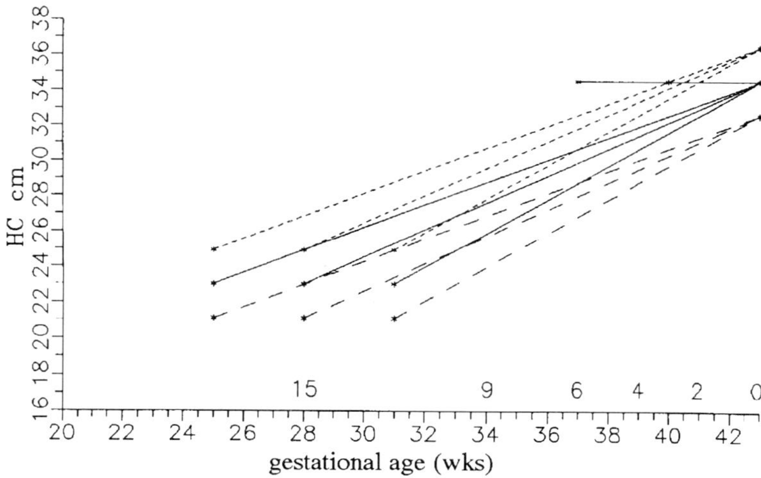

Figure 7.7 Prediction of confinement using biological–calendar scale for HC measurements

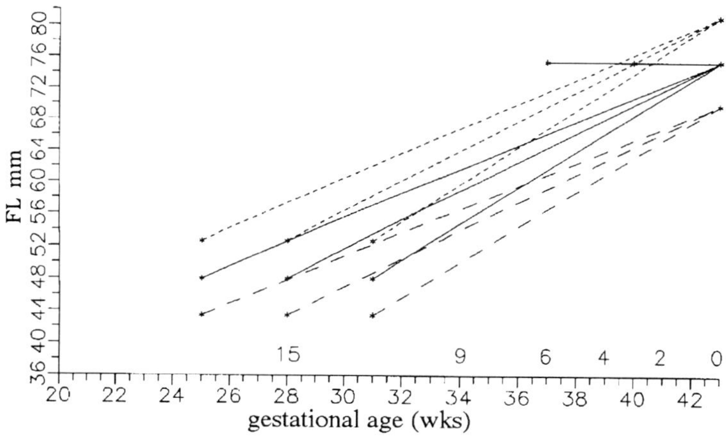

Figure 7.8 Prediction of confinement using biological–calendar scale for FL measurements

and ultrasonographic data and in future also with other imaging techniques. What is more, if the regression equations are known it is also possible to use the simple formula: $y = 2x + 22$ for the average values of oxytocinase increase after 25 weeks of pregnancy. This rule states that by doubling the oxytocinase level

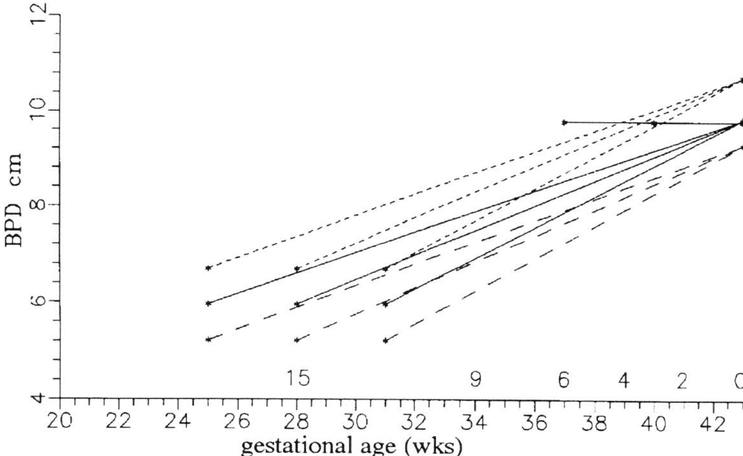

Figure 7.9 Prediction of confinement using biological–calendar scale for BPD measurements

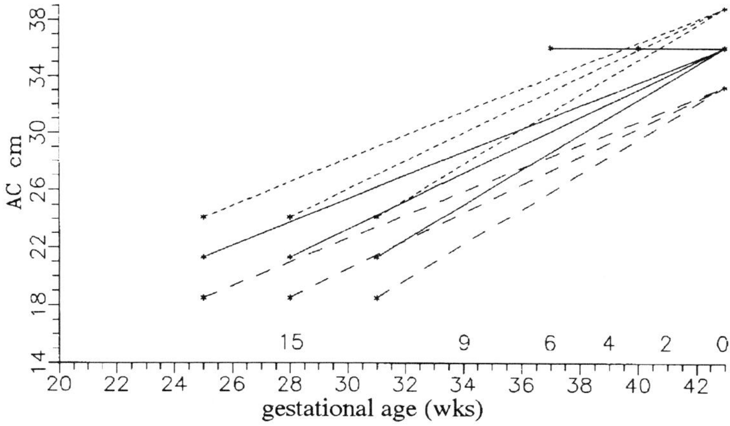

Figure 7.10 Prediction of confinement using biological–calendar scale for AC measurements

and adding 22 we obtain the enzymatically determined week of pregnancy duration. For example, the oxytocinase level of 5.0 IU (x) put into the equation indicates the 32rd week (y) of pregnancy. Using this rule, a doctor performing the sonographic measurements of the child is aware of the possible deviations of the child's biological development. On the other hand, being aware of the values of ultrasonographic measurements, it is possible to evaluate the independent enzymatic results in order to recognize or exclude the influence of the mother's neuro-hormonal condition.

In summary the constant increase in the oxytocinase level, and sonographic measurements, indicates the child's appropriate development. Their levels and direction enables the prediction of probable birth term with all the more accuracy, the more similar the subsequent values of the enzymatic and sonographic determinations. The gradient of the change in the values on the biological scale of pregnancy age is the decisive one.

COMPUTERIZED PREDICTION OF FETAL BIRTH-WEIGHT AND BIRTH-DATE

Progress has recently been made in implementing computerized techniques that can be used to obtain, for example, magnetic resonance images (MRI) in a fraction of a second rather than in minutes[26,27,171–173]. Echoplanar imaging (EPI) uses only one nuclear spin excitation per image and lends itself to a variety of critical medical applications, such as movie imaging of the mobile fetus *in utero*. What is more, MR images can reflect almost immediately a multitude of parameters, in contrast to the spatial distribution of a single parameter provided by acoustic (ultrasound), electron density (computed tomography), or isotope density (nuclear scanning) modes. Thanks to MRI the planar square and cubic dimensions used in the description of fetal development no longer needed to be reduced to linear functions. Especially this is important since *in utero* development is subject to periods of growth acceleration alternating with slower growth phases, e.g. in cases of neuroendocrinological gestosis.

Nowadays by completing an individual fetal growth profile (regular, fast and slow) rather than obtaining cross-sectional ultrasonographic data alone, the obstetrician is in a better position not only to detect abnormalities of fetal development; but also to monitor the outcomes of eventual therapeutic interventions and to predict the optimal date of childbirth. The incorporation of several fetal measurements into a composite assessment of calendar and biological gestational age generally enhances the accuracy of the computerized birth prediction. The presented computerized model takes into account independently all predictors of the gestational state and enables their simultaneous evaluation. In order to present obstetric

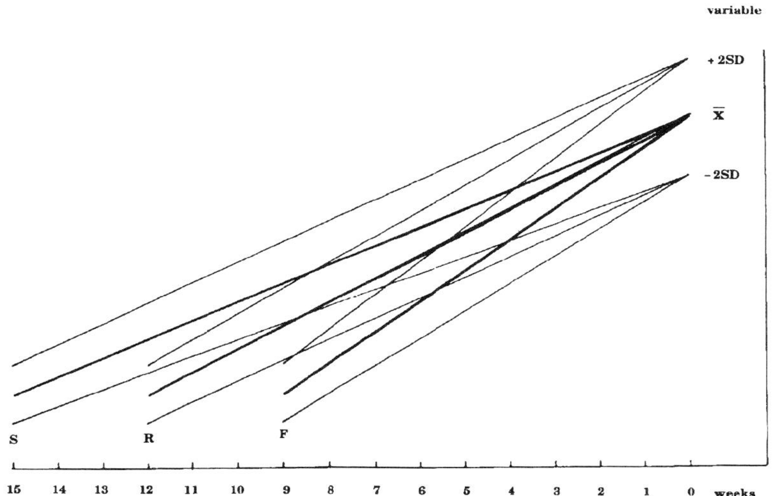

Figure 7.10a Slow (S), regular (R) and fast (F) fetal maturation to biological gestational age

measurements in a convenient format the different graphs can be obtained also as an instant on-screen display of the standard graphs of BPD, FL, AC and HC/AC ratio plotted against gestational age[12]. An apparatus for determining the optimum date of birth in different groups of high-risk pregnancies is under further development in cooperation with Siemens Medical Engineering Group.

Healthy pregnant women need only standard birth assistance, since so-called 'birth-medicine' seems inevitable only in less than 10% of births. Nevertheless, there are many complications involved with pregnancy and childbirth. Many of these disturbances present as increased risk during labor and birth, and therefore confirm the need for birth weight prediction.

Previously the basic problem in establishing connections between prenatal measurements and fetal weight was that weight could only be measured after birth. A new computerized clinical method for estimating intrauterine fetal weight and predicting birth-weight by use of ultrasound and biochemical measurements in the third trimester is presented, even without taking into account calendar gestational age at the time of examination. Estimation is based upon a computerized program for prediction of individual birth-date, automatically differentiating fast, regular and slow growing fetuses as well as indicating

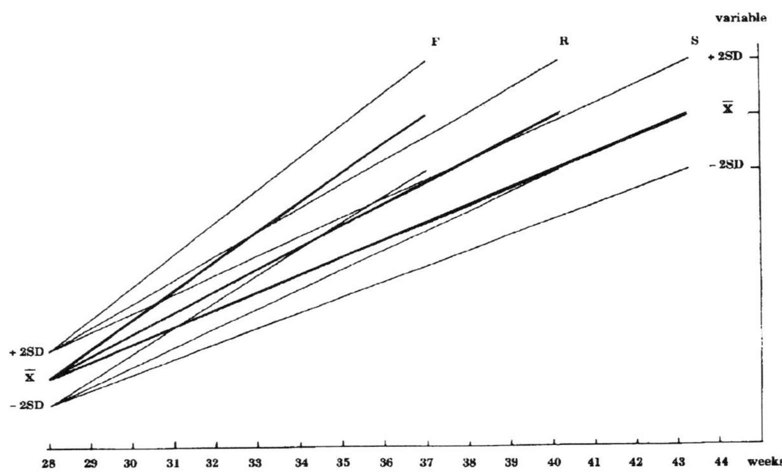

Figure 7.10b Fast (F), regular (R) and slow (S) fetal maturation to calendar gestational age

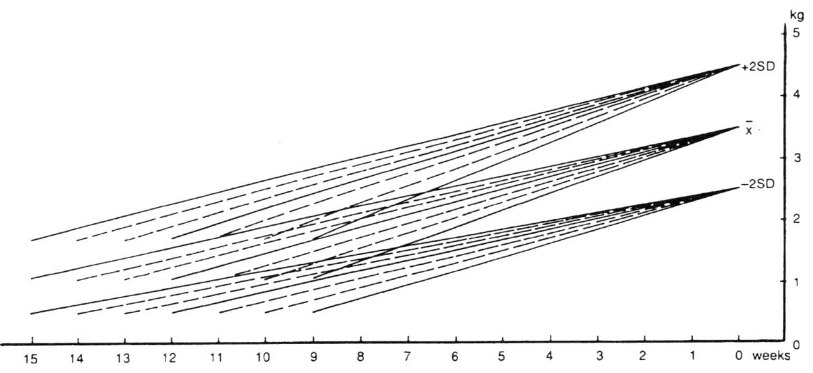

Figure 7.11a Fetal weight to biological age

physiological and pathological development of the individual pregnancy[7,8,12,13].

For a long time it has been known that average birth weight, fetal length and gestational length, by consecutive weeks or months of the year are identical. Everybody may compare this observation with any data collected within consecutive weeks or months at his or her own department. It is sufficient to encompass no more than two or three hundred single normal

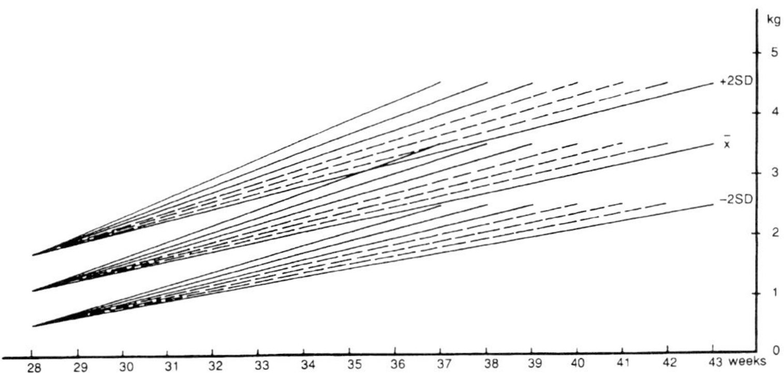

Figure 7.11b Fetal weight to calendar age

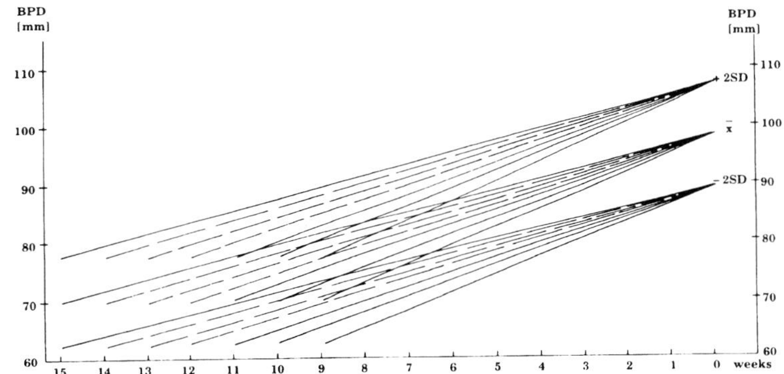

Figure 7.12a BPD to biological age

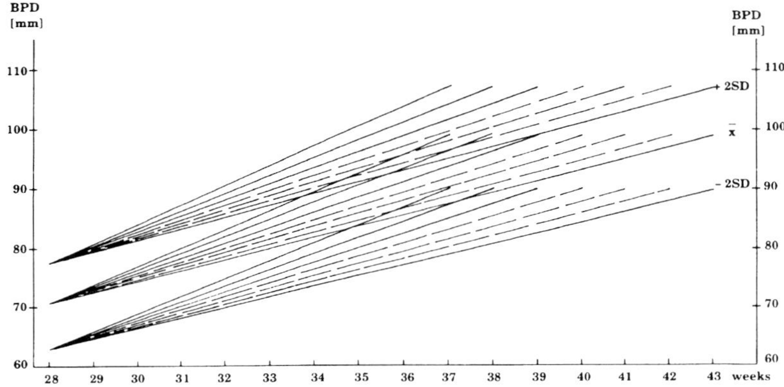

Figure 7.12b BPD to calendar age

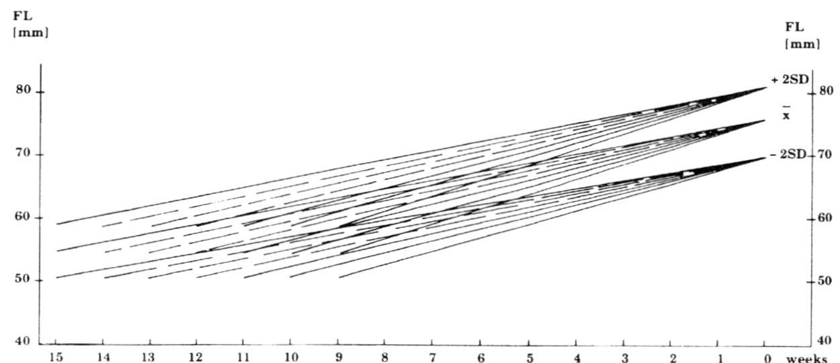

Figure 7.13a FL to biological age

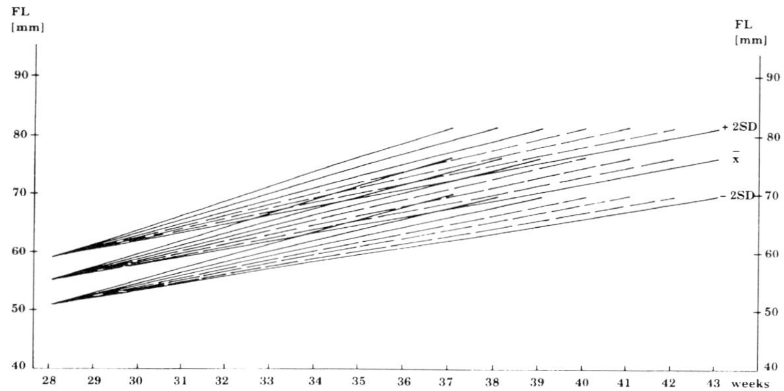

Figure 7.13b FL to calendar age

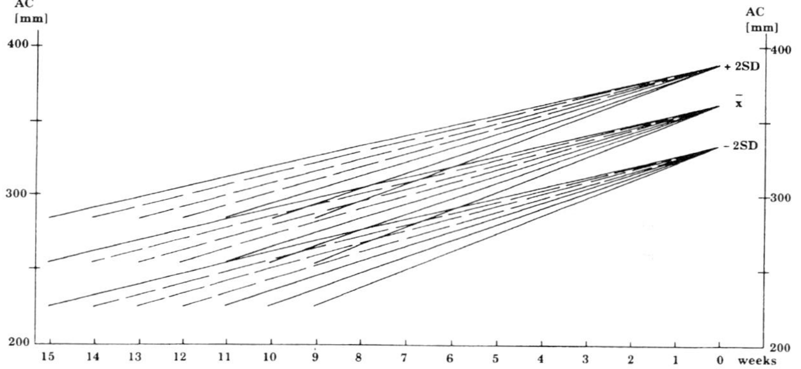

Figure 7.14a AC to biological age

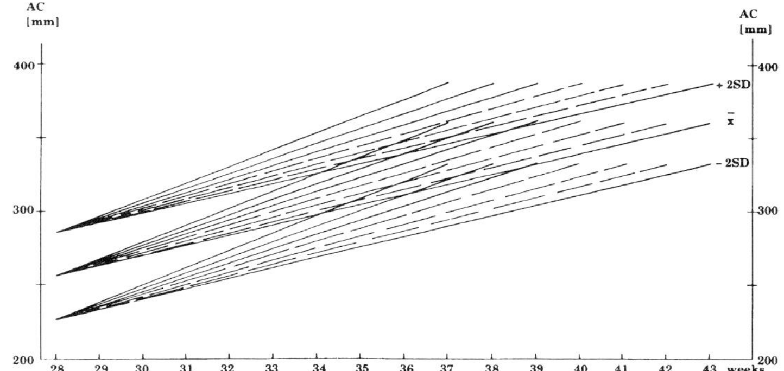

Figure 7.14b AC to calendar age

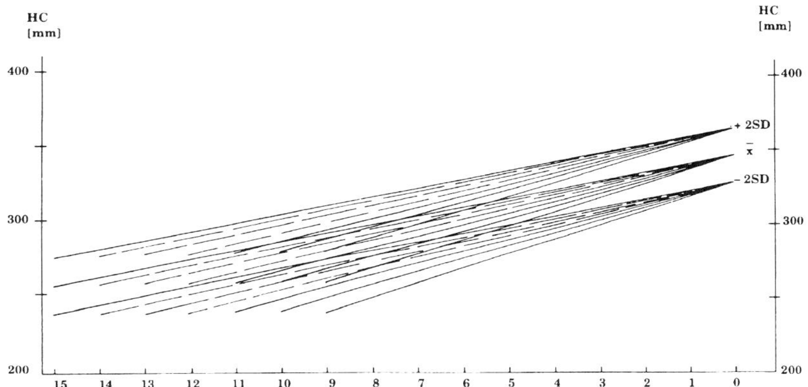

Figure 17.5a HC to biological age

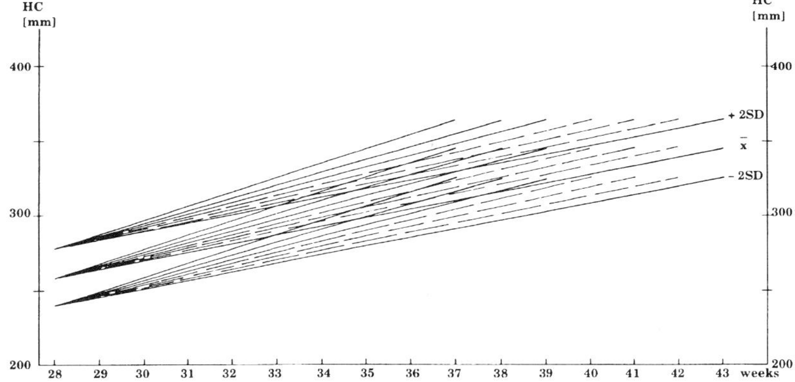

Figure 7.15b HC to calendar age

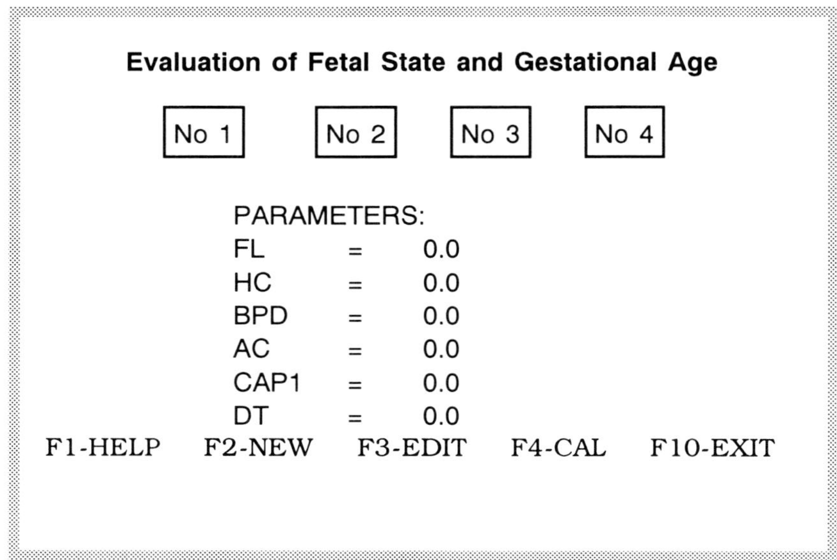

Figure 7.16 Computer screen

deliveries. Having such data one has to simply substitute birth mean values and their standard deviations with one's own data from $37^{0/7}$ to $43^{2/7}$ weeks, in the universal diagram for prediction of birth-date (Figure 7.10a,b) and then to replace also the mean and its ±2 standard deviations at the 28th week, with cross-sectional results of measurements or with former birth values three times smaller for weight or only decreased by one third for other ultrasound measurements[12].

The universal diagram (Figure 7.10a,b) may serve for preparing biological (Figure 7.10a) and calendar scales (Figure 7.10b) for fetal weight evaluation (Figure 7.11b) and for all ultrasonographic parameters. It is sufficient to collect the appropriate cross-sectional maternal parameters at 28 to 30 weeks' gestation and also within the last few days before the onset of spontaneous delivery of a single, healthy newborn. Then, appropriate scales for BPD, AC, HC and FL, are generated as in Figures 7.12–7.15.

The degree of fetal maturity and prediction of birth-date and weight by use of ultrasound may be assumed, if one of the following criteria is met:

(1) Two ultrasound measurements are performed within the interval of at least 15 days (optimum 4–5 weeks) starting ≥ 28 calendar gestational weeks, or

			PARAMETERS			
DATA	FL	HC	BPD	AC	CAP1	DT
1	58.0	268.0	76.0	242.0	4.2	0.0
2	69.0	296.0	86.0	290.0	8.0	5.0
3	76.0	350.0	96.0	356.0	10.2	5.0
4	0.0	0.0	0.0	0.0	0.0	0.0

DATA	CALENDAR AGE (weeks)	WEEKS TO LABOR	MATURITY TYPE	SCORE B&K	BIRTH B&K	ACTUAL WEIGHT (g)	BIRTH WEIGHT (g)
2	34.3	5.4	R	25.9	39.7	2340.9	3435.9
3	39.3	0.5	R	38.6	39.9	3357.3	3463.8

Figure 7.17 Data for patient U.K.

(2) One of the measurements was obtained earlier but in this case the date of the last menstrual period (LMP) is known.

Fetal mass or volume and gestational age can be calculated from AC, BPD, HC and FL independently or from any of them in such combination as will take into account only correct measurements, with the same accuracy of birth prediction.

When using the computerized pregnancy monitoring program, now patented by Siemens (P-293602), the computer screen (Figure 7.16) will request the following:

For each of the four consecutive examinations (No 1 to No 4), the physician must input requested fetal parameters and the time interval in weeks between consecutive measurements (DT). Some ultrasonographic equipment, for example Siemens, automatically registers the required ultrasound parameters. To illustrate this method, the following examples of actual computer print outs of examinations from pregnant women in our institute are presented (Figures 7.17–7.20). In each case the fetal parameters are shown from consecutive examinations (data 1 to 4) with the corresponding computer results.

In the first example (Figure 7.17), patient U.K., age 31, shows the following results:

The computer has calculated the calendar–gestational age based on the parameters for the second exam to be 34.3 and the third 39.3. In fact, the first exam was carried out at the

PARAMETERS

DATA	FL	HC	BPD	AC	CAP1*	DT
1	59.0	282.0	79.0	262.0	4.6	0.0
2	72.0	331.0	93.0	332.0	6.6	7.0
3	77.0	343.0	97.0	350.0	7.4	2.0
4	0.0	0.0	0.0	0.0	0.0	0.0

DATA	CALENDAR AGE (weeks)	WEEKS TO LABOR	MATURITY TYPE	SCORE B&K	BIRTH B&K	ACTUAL WEIGHT (g)	BIRTH WEIGHT (g)
2	37.8	2.4	R	34.1	39.9	3018.2	3484.7
3	39.8	1.0	R	38.8	41.1	3393.3	3581.9

Figure 7.18 Data for patient D.K.

PARAMETERS

DATA	FL*	HC*	BPD*	AC	CAP1	DT
1	58.0	287.0	78.0	0.0	2.8	0.0
2	60.0	285.0	81.0	0.0	3.8	2.0
3	62.0	300.0	84.0	0.0	4.8	3.0
4	0.0	0.0	0.0	0.0	0.0	0.0

DATA	CALENDAR AGE (weeks)	WEEKS TO LABOR	MATURITY TYPE	SCORE B&K	BIRTH B&K	ACTUAL WEIGHT (g)	BIRTH WEIGHT (g)
2	29.2	10.8	R	13.0	10.0	1340.0	3500.0
3	32.2	13.	M	18.0	41.0	1740.0	3580.0

Figure 7.19 Data for patient M.R.

30th week, the second at the 35th week and the third at the
40th week of calendar pregnancy duration. Furthermore, the
results indicate regular (R) fetal development with actual weeks
to delivery (5.4 and 0.5, respectively); neuromuscular maturation
scores (25.9 and 38.6 respectively); weight (2340.9 and 3357.3 g,

PARAMETERS						
DATA	FL*	HC*	BPD*	AC	CAP1*	DT
1	59.0	307.0	84.0	0.0	4.2	0.0
2	65.0	317.0	89.0	0.0	6.2	5.0
3	69.0	312.0	90.0	0.0	6.3	4.0
4	0.0	0.0	0.0	0.0	0.0	0.0

DATA	CALENDAR AGE (weeks)	WEEKS TO LABOR	MATURITY TYPE	SCORE B&K	BIRTH B&K	ACTUAL WEIGHT (g)	BIRTH WEIGHT (g)
2	36.2	8.1	M	25.0	40.2	2342.9	3545.8
3	40.2	?????	???	0.0	0.0	0.0	0.0

Figure 7.20 Data for patient A.S.

respectively) as well as the predicted birth weight and Ballard–Klimek scores.

In another example, patient D.K., age 23, has results shown in Figure 7.18. In this case, the computer additionally indicates an abnormal CAP_1 marked by 'flashing' on screen (note[*]). Thus, this parameter was not taken into consideration in calculating the fetal state. This should be a signal for the physician that something may be wrong in the course of this pregnancy or in oxytocinase determinations.

It must be pointed out, that in the physiological pregnancy, 90% of cases had computed results that conform almost completely with clinical assessment. In just 10% of cases with the help of this computerized method, one may uncover the abnormally developing pregnancy, with an indication of which parameters are altered or improperly determined.

This is illustrated by the next example (Figure 7.19), from patient M.R., age 31. In this case of neuroendocrinological gestosis treated by ACTH, only CAP_1 is used by the computer in calculating the fetal state (M-postmaturity), since 'flashing' parameters are not taken into consideration (FL, HC and BPD). However, spontaneous vaginal delivery took place at 38 weeks, with a fetal birth-weight of 2870 g and fetal length 51 cm.

In the last example (Figure 7.20), from patient H.S., age 29, at the time of the second examination, the fetal state is indicated.

However, it is not possible to compute a successive evaluation of the pregnancy, since the third examination was performed incorrectly. In fact the mother delivered by Cesarian section at 41st week due to fetal distress, with a fetal birth-weight of 3200 g and fetal length 54 cm. She was also treated with ACTH owing to neuroendocrinological gestosis.

The essence of computed measurement is that it relies on the analysis of angular increase of the variables in question. Our computerized scales are derived from our clinical data as well as the general literature[1–11,13,15,21,26,29,30,34,35,40–46,52,71,72,82,83,84,131,159,171,174–176,179–184]. We have prospectively examined several hundred cases to compare the computer generated data with known length of pregnancy, as illustrated in the following example.

In a group of 181 gravid women (average age 29.7±4.5 years) with clinically documented last menstrual period at their 35.3±2.3 gestation weeks the second computerized measurement was performed. Without taking into consideration LMP the length of the pregnancies was estimated as 35.8±2.6 weeks, which is not statistically different from the documented calendar gestation age ($F = 1.3$, $t = 1.9$). The pregnant women in this study gave birth after 3.1±2.1 weeks, which also does not differ statistically from predicted biological age (weeks to labor) of 3.5±2.2 weeks ($F = 1.1$, $t = 1.76$). The average newborn body weight was 3225±475 g, length 53.4±2.8 cm and Apgar score 9.7±0.7. A high level of correlation was shown between true and calculated calendar gestational age ($r = 0.93$, $t = 29$, $p < 0.001$) as well as between true and predicted birth-date ($r = 0.66$, $t = 10$, $p < 0.001$) although the increasing gestational oxytocinase profile, signifying physiological development of pregnancy was given only in 50.6% of cases.

The same results were obtained in another group of 47 women treated before becoming pregnant owing to neuroendocrinological disturbances. Again LMP gestation of 35.1±2.2 weeks was very similar to the computer estimated length of pregnancies: 35.7±2.8 ($F = 1.3$, $t = 1.1$, NS) as well as the calendar predelivery period (3±2.1 weeks) and the computer predicted value (3.6±2.3 weeks; $F = 1.1$, $t = 1.3$, NS).

The method presented here, based upon computerized data, is only beginning to open new clinical possibilities and should be perfected in the near future. Even now, however, it should oblige clinicians to change their evaluations of the relative duration of human pregnancy, especially when treating infants who will be born before and after the mean length of the gestation period. We hope it may contribute to lowering the mortality rates at both limits of normal birth as well as the incidence of premature deliveries.

The same has to be done by the commercial companies which in the advertisement for their different products, so often use the gestational calendar with incorrect obstetric information regarding the birth term and its normal range. The gestational calculator, described below, not only documents the clinical value of relative duration of pregnancy, but also can be used as an alternative to computer facilities. Above all, it should be used by the physician to properly inform the woman about her pregnancy.

GESTATIONAL CALCULATOR

The gestational calculator of biological and calendar age of pregnancy presents a means of predicting birth-date (term) using ultrasound and/or biochemical parameters of fetal maturity. The BPD, FL, HC, AC and CAP are measured at least twice in the last trimester of the gestation. It is worthwhile to underline that only 4% of births occur at a mean length of human pregnancy, 66% take place within ±11 days and 95% (considered as the biological norm) within ±3 weeks.

Based upon serial values, obtained several weeks apart (at least two, optimal 4–5 weeks), the fetus is determined to be growing at a slow, fast or regular rate. The corresponding weeks

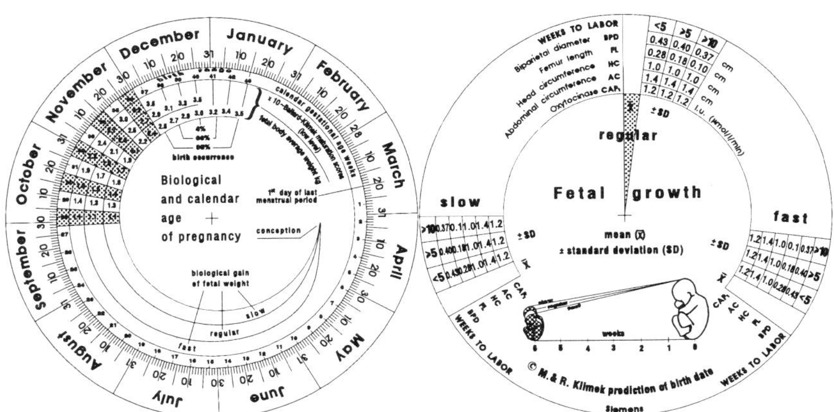

Figure 7.21 M. & R. Klimek© gestational calculator

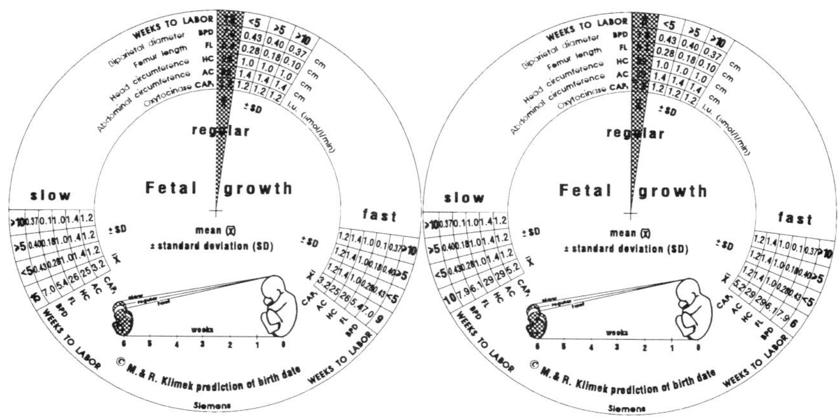

Figure 7.22 M. & R. Klimek© gestational calculator

to labor or 'term date' are recorded. This offers an accurate prediction of the birth-date as well as weight and lower limit of Ballard-Klimek score at examination (Figure 7.21).

INSTRUCTIONS FOR USE

Perform the appropriate ultrasonographic and biochemical measurements of the fetal parameters of maturity: BPD, FL, HC, AC, CAP and record these values, e.g. 7.0, 5.4, 26, 25 and 3.2, respectively.

In each of the three maturational profile windows (slow, regular and fast – Figure 7.21 right), find the matching values of these parameters and record the corresponding weeks to labor, i.e. 15, 12 and 9 weeks to labor, respectively. The estimates obtained of the weeks to labor are baseline measurements for each of the three possible maturational rates (Figure 7.22 left).

Perform the next measurements after, e.g. 4 weeks and evaluate in the same manner as before. Record the new values, e.g. 7.9, 6.1, 29, 29 and 5.2, respectively (Figure 7.22 right) and the corresponding weeks to labor, e.g. slow, 10; regular, 8; and fast, 6.

Next subtract the repeat values from the baseline estimates within each maturational profile (values from Figure 7.22 left minus Figure 7.22 right).

From among the three maturational profile windows the difference in weeks that most closely matches the elapsed weeks

(i.e. 4) between the two serial measurements:
- slow 15 – 10 = 5
- regular 12 – 8 = 4

and
- fast 9 – 6 = 3

gives the appropriate growth rate as regular and corresponding 8 weeks to labor and predicted birth-date (date at last examination + 8 weeks).

8
Conclusions

Rudolf Klimek

The enzymatic and ultrasonographic monitoring of pregnancy and prediction of the date and course of delivery constitutes one of the lasting achievements of contemporary obstetrics. It means that pregnancy is regarded not only in terms of the state of the maternal organism but as the coexistence of at least two human beings in gestational unity.

Oxytocinase and its isoenzymes reflect the present state of mother, infant and placenta, all three being inseparable components of the pregnancy until labor. The oxytocin–oxytocinase system symbolizes this unity, enabling evaluation and active intervention in the basic informative processes referring to its neurohormonal regulation. What is more, examination of the equilibrium of the oxytocin–oxytocinase system is available in almost every obstetric ward. That is why neglect of basic clinical examinations starting from at least the onset of the 37th week of pregnancy on the one hand, and the induction of labor only on the basis of calendar gestational age on the other hand, are unacceptable from an ethical point of view.

Medicine has to use the proper methods of distinguishing the baby which should be born at the beginning of the normal range from that which should be born at the end of the normal range, or between these limits. In some cases medical intervention could be found to be irresponsible if the parents asked the doctor if he had done everything that should have been done from the diagnostic, but not exclusively statistical, point of view.

The fact of the matter is that medicine, unlike any other field of human activity, or – regrettably – also lack of activity in some instances, requires from the physicians the specific responsibility for conscientious and continued acquisition of knowledge for the good of patients. Certainly we expect a proper ethical and moral attitude from those who are responsible for the appropriate care of mother and child, and for progress in maternal–fetal medicine in accordance with the development of general human knowledge. In making this book available to all physicians we not only want to fulfill our duty to inform

[75]

and educate but also to change the negative attitudes of those who do not yet understand obstetric enzymology and/or misuse sonography. This is because we regard with equal accountability the old medical rule *primum non nocere* and our duty as physicians to act for the good of our fellow man.

9
References

1. Klimek R. XI miesiac ksiezycowy ciazy jako istotne pojecie poloznicze. *Pol. Tyg. Lek.*, 1964; 19: 109.
2. Klimek R. Pregnancy and labor in terms of studies on the oxytocin–oxytocinase system. *Folia Med. Cracoviensis*, 1964; 6: 471; and In: *Oxytocin and its Analogous.* Klimek R., Król W. (eds.) PTE Cracow 1964, p. 66
3. Klimek R. Relative duration of human pregnancy and oxytocin therapy. Part I. *Gynaecologia*, 1967; 163: 48.
4. Klimek R. Relative duration of human pregnancy and oxytocin therapy. Part II Enzymic block. *Gynaecologia*, 1967; 163: 54.
5. Klimek R. The oxytocin–oxytocinase system as a model of mother–foetus relations. Intra-uterine danger to the foetus. *Exc. Med. Congress* Monograph. Amsterdam, 1967, p. 504.
6. Klimek M. *Enzymatic and Ultrasonografic Prediction of Delivery Term.* Doctor's Thesis. AM Cracow, 1991
7. Klimek M. Prediction of birth-date and birth-weight based upon the percentile distribution of ultrasound cross-sectional data. *Ginekologia Polska*, 1992; 63: 400.
8. Klimek M., Elaccari S., Klimek M., Tomaszczyk B. Computerized confirmation of calendar gestational age. *Ginekologia Polska*, 1992; 63: 500.
9. Klimek M., Klimek R., Michalski A. The effectiveness of oxytocinase determinations to terminate pregnancy. XIII World Congress of FIGO Singapore. *Int. J. Obstet. Gynecol.*, 1991; 20: 1324 and In: Klimek R. (ed.) *Pre- and Perinatal Psycho-medicine.* DWN DReAM, Cracow 1992, p. 9.
10. Klimek R.: Enzymatic and ultrasonographic monitoring of pregnancy and prediction of birth-date. DWN DReAM Cracow, 1992.
11. Klimek R. (ed.) *Pre- and Perinatal Psychomedicine.* DWN DReAM Cracow, 1992, p. 9.
12. Klimek R. Computerized prediction of fetal birth-weight and birth-date using ultrasound measurements in advanced pregnancy. *Ultrasound Obstet. Gynecol.*, 1992; 2: Suppl. 1: 119.
13. Klimek R., Fraczek A. Computerized prediction of fetal birth-weight for MRI evaluation of fetal maturity. *Int. J. Prenatal Perinatal Stud.*, 1992; 4: Suppl. 1, 112.

14. Bogdanowicz M., Dawidiuk W., Kalita J., Paradysz A., Reron A., Stanek J., Wachta T. Kliniczne i statystyczne uwagi o rozpoznawaniu ciazy przenoszonej. *Pamietnik XVIII Naukowego Zjazdu Ginekologów Polskich Bialystok*, 1971; pp. 187–190.

15. Creasy R.K., Resnik R. (eds.) *Maternal–Fetal Medicine: Principles and Practice*. W.B. Saunders Company, Philadelphia, 1989.

16. Hosemann H. Normale und abnormale Schwangerschaftsdauer. In: *Biologie und Pathologie des Weibes*. Verlag Urban Schwarzenberg, Berlin, 1952, p. 828.

17. Hummel K. *Die Medizinische Vaterschaftsbegutachtung mit Biostatischen Beweis*. Stuttgart, 1961.

18. Sabbagha R.E. *Diagnostic Ultrasound Applied to Obstetrics and Gynecology*. J.B. Lippincott & Company, Philadelphia, 1987.

19. Stave U. *Perinatal Physiology*. Plenum Medical Book Company, New York and London, 1970.

20. Stegmann H., Hellwig G. *Die medizinische Varenschuftbegutachtung mit biostatischen Bewers*. Stuttgart, 1961.

21. Turnbull Sir A., Chamberlain G. (eds.) *Obstetrics*. Churchill Livingstone, Edinburgh and London, 1989.

22. Yerushalmy J. Relation of birth weight, gestational age, and the rate of intrauterine growth to perinatal mortality. *Clin. Obstet. Gynecol.*, 1970; 13: 107.

23. Zdebski Z, red. Zasady postepowania w naglych stanach polozniczo-ginekologicznych. *PZWL*, Warszawa, 1991. wyd 2.

24. Dunn P.M. The Bristol perinatal growth chart: a simple flexible standard based on normal growth velocity. *Working Paper for WHO Consultation on Reporting and Analysis of Perinatal and Maternal Morbidity*, Bristol (25–27 September), 1972, pp. 1–40. Held in Files of ICD Unit, WHO, Geneva.

25. Belfort P., Pinotti J.A., Eskes T.K.A.B. (eds.) Pregnancy and labor. In: *Advances in Gynecology and Obstet*rics. Parthenon Publishing, New Jersey, 1988.

26. Klimek R. Magnetic resonance imaging and spectroscopy. In: Maeda K. *et al.* (eds.) *Computer and Perinatal Medicine*. Elsevier Science Publishers, 1990, p. 173

27. Klimek R. Introduction to the technology of magnetic resonance imaging in infertility. In: Mashiach S. *et al.* (eds.) *Advances in Assisted Reproductive Technologies*. Plenum Press, New York, 1990, pp. 525–531.

28. Dunn P.M. (ed.) *Report of the FIGO Sub-Committee on Perinatal Epidemiology*, Cairo, 1984.

29. Gruenwald P. The fetus prolonged pregnancy. *Am. J. Obstet. Gynecol.*, 1964; 89: 503–509.

30. Gruenwald P. Growth of the human fetus. I. Normal growth and its variations. *Am. J. Obstet. Gynecol.*, 1966; 94: 1112–9.

31. Lubchenco L., Hansman C., Dressler M., Boyd E. Intrauterine growth as estimated from liveborn birthweight data at 24 to 42 weeks of gestation. *Pediatrics*, 1963; 32: 793–800.

32. Thomson A.M., Billewicz W.Z., Hytten F.E. The assessment of fetal growth. *J. Obstet. Gynaecol. Br. Cwlth.*, 1968; 75: 903–16.

33. Klimek R. *Enzymatic Monitoring of Pregnancy.* EAGO PLB ISPPM Cracow, 1991

34. Klimek R., Michalski A., Milewicz S., Rzepecka A., Szlachcic M., El Accari S., Fraczek A. Results of psychohormonal prevention of premature deliveries. *Int. J. Prenatal Perinatal Stud.*, 1991; 3: 87–95.

35. Filipowicz A. Ultrasonograficzna weryfikacja biochemicznego monitorowania ciazy. Rozprawa doktorska AM Kraków, 1991.

36. *ACOG Committee Opinion: Fetal Maturity – Assessment Prior to Elective Repeat Cesarean Delivery,* No 98. September 1991

37. Benacerraf B.R., Gelman R., Frigoletto F. Sonographically estimated fetal weights: accuracy and limitation. *Am. J. Obstet. Gynecol.*, 1988; 159: 1118.

38. Chitkara U., Rosenberg J. *et al.* Prenatal sonographic assessment of the fetal thorax: normal values. *Am. J. Obstet. Gynecol.*, 1987; 156: 1069.

39. Diaferia A., D'Agostino G. *et al.* Bari district birth weight study: comparison with the Bristol perinatal growth chart. *Int. J. Gynecol. Obstet. Gynecol.*, 1989; 29: 227–231.

40. Fleischer A. (ed.) *The Principles and Practice of Ultrasonography in Obstetrics and Gynecology,* 4th edn. Appleton and Lange, New York, 1990.

41. Goldenberg R.L., Davis R.O. *et al.* Prematurity, postdates, and growth retardation: the infuence of use of ultrasonography on reported gestational age. *Am. J. Obstet. Gynecol.*, 1989; 160: 462.

42. Goldstein I., Reece E.A., Pill G. Cerebral measurements with ultrasonography in the evaluation of fetal growth and development. *Am. J. Obstet. Gynecol.*, 1987; 156: 1065.

43. Goldstein I., Reece E.A., Hobbins J.C. Sonographic appearance of the fetal heel ossification centers and foot length measurements provide independent markers for gestational age estimation. *Am. J. Obstet. Gynecol.*, 1988; 159: 923.

44. Hadlock F.P., Harrit R.B. *et al.* Estimating fetal age using multiple parameters: a prospective evaluation in a racially mixed population. *Am. J. Obstet. Gynecol.*, 1987; 156: 955.

45. Hensmann M., Hackelöer B. Ultraschaldiagnostic in der Geburtshilfe und Gynäkologie. Springer–Verlag, Berlin, 1984.

46. Hill L.M., Guzick D. *et al.* Composite assessment of gestational age: a comparison of institutionally derived and published regression equations. *Am. J. Obstet. Gynecol.*, 1992; 166, 2: 551

47. Kurjak A. (ed.) *Fetus as a Patient.* Excerpta Medica, Amsterdam, 1985

48. Kurjak A., Jurkovic D., Alfirevic Z. Fetal therapy – state of the art. *Int. J. Prenatal Perinatal Stud.*, 1989; 1: 21–46.

49. Kurjak A., Kossof G. (eds.) *Recent Advances in Ultrasound Diagnosis.* Excerpta Medica, Amsterdam, 1986.

50. Landon M.B., Mintz M.C., Gabbe S.G. Sonographic evaluation of fetal abdominal growth: predictor of the large-for-gestational-age infant in pregnancies complicated by diabetes mellitus. *Am. J. Obstet. Gynecol.*, 1989; 160: 115.

51. Mhaskar R., Agarwal N. *et al.* Fetal foot length – a new parameter for assessment of gestational age. *Int. J. Gynecol. Obstet.*, 1989; 29: 35–38.

52. Merz A. *Ultrasound in Gynecology and Obstetrics – Text and Atlas.* G. Thieme, New York, 1991.

53. Miller J.M., Brown H.L. *et al.* Ultrasonographic identification of the macrosomic fetus. *Am. J. Obstet. Gynecol.*, 1988; 159: 1110.

54. Ott W.J., Dayle S. Normal ultrasonic fetal weight curve. *Am. J. Obstet. Gynecol.*, 1982; 59: 603

55. Sack R.A., Makarry J.M. Misdiagnoses in obstetric and gynecologic ultrasound examinations: causes and possible solutions. *Am. J. Obstet. Gynecol.*, 1988; 158: 1260.

56. Sarmandal P., Grant J.M. Effectiveness of ultrasound determination of fetal abdominal circumference and fetal ponderal index in the diagnosis of asymmetrical growth retardation. *Br. J. Obstet. Gynaecol.*, 1990; 97: 118–123.

57. Spaczynski M., red. Ultrasonografia w poloznictwie i ginekologii. *PZWL*, Warszawa, 1989.

58. Tamura R.K., Sabbagha R.E., Dooley S.L. *et al.* Real time ultrasound estimations of weight in fetuses of diabetic gravid women. *Am. J. Obstet. Gynecol.*, 1985; 153: 57.

59. Berde B. (ed.) Neurohypophysical hormones and similar polypeptides. *Handbook of Experimental Pharmacology*, Vol XXIII. Springer–Verlag, Berlin–Heidelberg–New York, 1968.

60. Christensen A., Fryshols D., Fylling P. Hormone and enzyme assays in pregnancy. IV The human chorionic somatomammotrophin, placental cystine aminopeptidase, progesterone and the urinary oestrogens in

pregnancies complicated with essential hypertension, mild or severe pre-eclampsia. *Acta Endocrinol.*, 1974; 77: 344.

61. Page E.W., Titus M.A., Mohun G., Glendening M.B. The origin and distribution of oxytocinase. *Am. J. Obstet. Gynecol.*, 1961; 82: 1090–1095.

62. Ryden G. Cystine aminopeptidase activity in pregnancy. *Acta Obstet. Gynecol. Scand.*, 1971; 50: 253–257.

63. Semm K. Die klinische Bedeutung der Oxytocinase bestimmung. Klin. Wochenschr., 1955; 33: 817–818.

64. Tuppy H. The influence of enzymes on neurohypophysical hormones and similar peptides. In: Berde B. (ed.) *Neurohypophysical Hormones and Similar Polypeptides. Handbook of Experimental Pharmacology.* Springer–Verlag, Berlin–Heidelberg–New York, 1968; Vol XXIII: 66.

65. Nalepa J. Kliniczno-laboratoryjne badanie cystynoamino- peptydazy surowicy krwi kobiet ciezarnych. Rozprawa doktorska AM Kraków, 1980

66. Rzepecka A. Powtarzalnosc wyników chemicznej metody oznaczen oksytocynazy. Rozprawa doktorska AM Kraków, 1991.

67. Michalski A. Poród spontaniczny i prowokowany a profil oksytocynazemii w ciazy. Rozprawa doktorska AM Kraków, 1991.

68. Klimek R., Filipowicz A., Klimek M., Michalski A., Rzepecka A. Kliniczna przydatnosc przedporodowego oznaczania oksytocynazemii. *Gin. Pol.*, 1991; 62: 367.

69. Semm K. Serum-Oxytocinase-Aktivitat in der Placenta. *Arch. Gynakol.*, 1963; 199: 265–270.

70. Semm K. Oxytocin: abbau durch Enzyme. *Gynaecologia (Basel)*, 1965; 159: 61–63.

71. Klimek R. Clinical studies on the balance between isooxytocinases in the blood of pregnant women. *Clin. Chim. Acta*, 1968; 20: 235.

72. Klimek R. Enzyme monitoring of fetal development. *Proceedings of the VII World Congress of Obstetrics and Gynecology*, Moscow, 1973, p. 196.

73. Smyth C.N. Uterine irritability; the oxytocin sensitivity test. *Lancet*, 1958; 1: 237.

74. Klimek R., Bieniasz A. Studies on the relation between serum oxytocinase and labor. *Am. J. Obstet. Gynecol.*, 1969; 104: 959.

75. Klimek R., Bieniasz A., Drewniak K. Further studies on the oxytocin–oxytocinase system. *Am. J. Obstet. Gynecol.*, 1969; 105: 427.

76. Klimek R., Bieniasz A., Drewniak K., Zdebski Z. Reaktywnosc macicy kobiety ciezarnej na oksytocyne a reguly oznaczania terminu porodu przy uzyciu testu oksytocynowego. *Gin. Pol.*, 1970; 41: 625.

77. Klimek R., Madej J., Pietrzycka M. Wartosc kliniczna dozylnego testu oksytocynowego. *Pol. Tyg. Lek.*, 1961; 16: 1309.

78. Klimek R., Malolepszy E. Evaluation of a rapid enzymatic method for determining placental function. *Clin. Chim. Acta*, 1969; 24: 349.

79. Klimek R., Stanek J. O koniecznosci oznaczania ukladów substratenzym w diagnostyce polozniczej. *Gin. Pol.*, 1973; 4: 445–453.

80. Klimek R., Zdebski Z. Oxytocinogenic activity of hormonogen triglycylooxytocin in the human body in relation to the oxytocin–oxytocinase system. *Eur. J. Clin. Invest.*, 1974; 4 386.

81. Drewniak K. Clinical evaluation of the biochemical method of determining the term of delivery. In: Klimek R., Król W. (eds.) *Oxytocin and its Analogues.* Kraków: Ed. Polish Endocrinol. Soc., 1964, pp. 55–58.

82. Krzyczkowska-Sendrakowska M., Kaim I., Klimek R., Klimek M., Rzepecka A. Przedporodowa izooksytocynazemia u kobiet ciezarnych z cukrzyca lub nadcisnieniem. *Gin. Pol.*, 1991; 62: 421.

83. Klimek R. Les resultates therapeutiques dans des cas du syndrome hypothalamique post-gravidique. *Actual. Endocrinol.*, 1965; 9: 195.

84. Klimek R. Diagnoza a terapie ohrozenych tehotenstvi pomoci syntetickeho ACTH. *Cs. Gynecol.*, 1973; 38: 647

85. Klimek R., Michalski A., Milewicz S. *et al.* Results of psychohormonal prevention of premature deliveries. *Int. J. Prenatal Perinatal Stud.*, 1991; 3: 87–95.

86. Klimek R., Milewicz S., Paradysz A., Stanek J., Walas E. Postepowanie i wyniki leczenia w ciazy endokrynologicznie zagrozonej. Postepy ginekologii i poloznictwa. *XIX Zjazd PTG Katowice*, 1974; pp. 42–44.

87. Klimek R., Skolicki Z. Wyniki leczenia ciazy zagrozonej przy uzyciu syntetycznego ACTH. *Gin. Pol.*, 1979; Suppl., 254

88. Skolicki Z. Ocena leczenia hormonalnego ciazy zagrozonej. Rozprawa doktorska AM Kraków, 1977.

89. Klimek R., Kallista-Milewicz W., Skolicki Z. Evaluation and follow-up studies of newborns whose mothers received synthetic ACTH in pregnancy. *Eur. Soc. Clin. Invest.*, 1980; 10: 2

90. Klimek R., Król W. (eds.) *Oxytocin and its Analogues.* Cracow PTE, 1964.

91. Klimek R., Krzysiek J., Paradysz A., Stanek J. Blood oxytocinases and phosphatases levels in pregnant women treated with synacthendepot. *Pol. Endocrinol.*, 1978; 2: 121.

92. Klimek R., Milewicz S. Reduction of RDS mortality in newborns after administration of synthetic ACTH to the mothers. *Gin. Pol.*, 1983; 5: 341.

93. Krzysiek J., Billewicz O., Klimek R. Tomografia komputerowa przysadki mózgowej po ciazy powiklanej gestoza. *Pam. V Symp. Sekcji Gestoz PTG Szczecin*, 1985; 276

94. Lauritzen C.A. Clinical test for placental functional activity using DHEA sulphate and ACTH injections in the pregnant woman. *Acta Endocrinol.*, 1967; Suppl. 119: 188.

95. Reck G., Noss V., Breckwoldt M. Response of plasma non-conjugated oestradiol to manipulated adrenal function by ACTH and dexamethasone. *Acta Endocrinol. (DK)*, 1979; 3: 525.

96. Rees L.H., Burke C.W., Chard T. *et al.* Possible placental origin of ACTH in normal human pregnancy. *Am. J. Obstet. Gynecol.*, 1975; 254: 620.

97. Carr B.R., Parker C.R. *et al.* Maternal plasma adrenocorticotropin and cortisol relationships throughout human pregnancy. *Am. J. Obstet. Gynecol.*, 1981; 146: 416.

98. Conway D.I, Anderson D.C., Gordon M.T. *et al.* Adrenal progesterone, 17 alpha-hydroxyprogesterone and cortisol responses to synacthen in normal women and women with various gynecological disorders. *Clin. Endocrinol.*, 1983; 1: 77.

99. Curet L.B., Morrison J.C. *et al.* Antenatal therapy with corticosteroids and postpartum complications. *Am. J. Obstet. Gynecol.*, 1985; 152: 83.

100. Genazzini A.R. *et al.* Immunoreactive ACTH and cortisol plasma levels during pregnancy. Detection and partial purification of corticotrophin like placental hormone: the human chorionic corticotrophin. *Clin. Endocrinol.*, 1975; 41.

101. Goland R.S., Wardlaw S.L. *et al.* Biologically active corticotropin-releasing hormone in maternal and fetal plasma during pregnancy. *Am. J. Obstet. Gynecol.*, 1988; 159: 884.

102. Klimek R., Stanek J. Ciaza o wysokim ryzyku w ujeciu neuroendokrynologicznym. *Gin. Pol.*, 1976; 47: 89–100.

103. Laatikainen T.J., Raisanen I.J., Salminen K.R. Corticotropin-releasing hormone in amniotic fluid during gestation and labor and in relation to fetal lung maturation. *Am. J. Obstet. Gynecol.*, 1988; 159: 891.

104. Liggins G.C., Howie R.N. A controlled trial of antepartum glucocorticoid treatment for prevention of respiratory distress syndrome in premature infants. *Pediatrics*, 1972; 50: 515.

105. Liggins GC. Adrenocortical-related maturational events in the fetus. *Am. J. Obstet. Gynecol.*, 1976; 126: 931.

106. Lohmeyer H. Kontraindikationen und Nebenwirkungen der Cortisontherapie in der Schwangerschaft. *Zbl. Gynak.*, 1963; 85: 833.

107. Milewicz-Kallista W. Rozwój dzieci matek leczonych Synacthen-Depot w okresie ciazy. Rozprawa doktorska AM Kraków, 1977.

108. Petraglia F., Sutton S., Vale W. Neurotransmitters and peptides modulate the release of immunorective corticotropin-releasing factor from cultured human placental cells. *Am. J. Obstet. Gynecol.*, 1989; 160: 247.

109. Reck G., Metzger M. *et al.* Suppresion des freien Plasmaoestriols bei der Risikoschwangerschaft nach Stimulation der Muetterlichen Nebennierenrinde mit ACTH. *Geburtsh Frauenheilk.*, 1980; 6: 508.

110. Romanini C., Valensise H., Ardvini D. Feto-maternal interactions in normal and pathological pregnancies. In: Cosmi E.V. (ed.) *Proceedings of XI European Congress of Perinatal Medicine.* Hardwood Academic Publishers, Switzerland, 1989.

111. Simmer H.H., Tulchinsky D. *et al.* On the regulation of estrogen production by cortisol and ACTH Regulation in human pregnancy at term. *Am. J. Obstet. Gynecol.*, 1974; 119: 283.

112. Wilson E.A., Jawad M.J. Stimulation of human chorionic gonadotropin secretion by glucocorticoids. *Am. J. Obstet. Gynecol.*, 1982; 142: 344.

113. Zsolnai B., Gyevai A., Troszynski M. Induction of lamellar bodies synthesis by corticosteroids in first trimester human fetal lungs. *Gin. Pol.*, 1984; 6: 385.

114. Zuidema L.J., Khan-Dawood F. *et al.* Hormones and cervical ripening: dehydroepiandrosterone sulfate, estradiol, estriol, and progesterone. *Am. J. Obstet. Gynecol.*, 1986; 155: 1252.

115. Szlachcic M. Profile izooksytocynazemii u kobiet ciezarnych leczonych adrenokortykotropina. Rozprawa doktorska AM Kraków 1991.

116. Page E.W. The value of plasma pitocinase determinations in obstetrics. *Am. J. Obstet. Gynecol.*, 1946; 52: 1014–1022.

117. Tuppy H., Nesvadba H. Uber die Aminopeptidaseaktivitat des Schwangerenserums und ihre Beziehung zu dessen Vermogen, Oxytocin zu inaktivieren. *Mh. Chem.*, 1957; 88: 977–988

118. Jacyszyn K., Ruchlewicz B., Wisniewska M., Walas J. Aktywnosc oksytocynazy, aminopeptydazy leucynowej oraz transferazy glutamylowej pod koniec trwania ciazy prawidlowej oraz w ciazy przeterminowanej. *Gin. Pol.*, 1977; 48: 1043.

119. Kleiner H., Dictus-Vermeulen C., May-Cocriamont C. *et al.* Human placental oxytocinase and relationship to pregnancy plasma oxytocinase. *Clin. Chim. Acta*, 1980; 101: 113–123.

120. Mizutani S., Yoshino M., Oya M. Placental and nonplacental leucine aminopeptideses during normal pregnancy. *Clin. Biochem.*, 1976; 16–18.

121. Mizutani S., Yoshino M., Oya M. A comparision of oxytocinase and L-methionino-insensitive leucine aminopeptidase during normal pregnancy. *Clin. Biochem.*, 1976; 9: 228.

122. Nalepa J. Zarys badan nad oksytocynaza w poloznictwie. *Gin. Pol.*, 1986; 10: 669.

123. Nalepa J., Gadek K. Dalsza modyfikacja chemicznej metody Tuppy'ego i Nesvadby w modyfikacji Klimka oznaczania aktywnosci cystynoaminopeptydazowej (oksytocynazy) w surowicy krwi. *Gin. Pol.*, 1986; 9: 622.

124. Sciarra J.J., Burress D.A. Serum leucine aminopeptidase in twin pregnancy. Proc. Soc. Exp. Biol., 1960; 104: 712–713.

125. Beck L. *Zur Geschichte der Gynakologie und Geburtshilfe*. Springer–Verlag, Berlin, 1986

126. Bishop E.H. Pelvic scoring for elective induction. *Obstet. Gynecol.*, 1964; 24: 266–268.

127. Dale H.H., Dudley H.W. On the pituitary active principles and histamine. *J. Pharmacol. Exp. Ther.*, 1921; 18: 27–42.

128. Du Vigneaud V., Ressler C., Swan J. M., Roberts C. W., Katsoyannis P.G. The synthesis of oxytocin. *J. Am. Chem. Soc.*, 1954; 76: 3115–3121.

129. Du Vigneaud V., Ressler C., Trippett S. The sequence of amino acids in oxytocin, with a proposal for the structure of oxytocin. *J. Biol. Chem.*, 1953; 205: 949–957.

130. Edebin A.A., Mack D.S., McDonald D.J., Philips J. The stability and short-term fluctuation in serum oxytocinase activity in pregnancy. *Int. J. Obstet. Gynecol.*, 1989; 28: 331–335.

131. Elaccari S. Accuracy of gestational age prediction by enzymatic and ultrasonographic measurements in neuro-endocrinological patients. *Ginekologia Polska*, 1992; 63: 700

132. Klimek R., Paradysz A. Enzymatic monitoring of pregnant women treated with synthetic ACTH because of threatened abortion. *Eur. J. Clin. Invest.*, 1973; 3: 243

133. Klimek R., Zdebski Z. The reactivity of the human uterus to desaminooxytocin in the prepartum period. *Pol. Med. Sci. Hist.*, 1968; 129.

134. Oya M., Yoshino M., Asano M. Human placental aminopeptidase isoenzymes. *Experientia*, 1974; 30: 985–986.

135. Oya M., Yoshino M., Mizutani S., Wakabayashi T. The origin of human pregnancy serum oxytocinase. *Gynecol. Invest.*, 1974; 5: 276–283.

136. Oya M, Yoshino M, Mazutani S. Molecular heterogeneity of human placental aminopeptidase isoenzymes. *Experientia*, 1975; 31: 1019–1020.

138. Traczyk W., Trzebski A., red. Fizjologia czlowieka. *PZWL*, Warszawa, 1980.

139. Tuppy H. The amino-acid sequence in oxytocin. *Biochim Biophys Acta*, Amsterdam 1953; 11: 449–450.

140. Tuppy H., Trippett S. The sequence of amino acids in oxytocin, with a proposal for the structure of oxytocin. *J. Biol. Chem.*, 1953; 205: 949–957.

141. Tuppy H.C., Ressler C., Swan J.M., Roberts C.W. *et al.* The synthesis of oxytocin. *J. Am. Chem. Soc.*, 1954; 76: 3115–3121.

142. ACOG Technical Bulletin: *Ultrasound in Pregnancy*, No 116. May 1988

143. Bao-Jan J., Kaufman R.H., Franklin R.R. Vaginal cytology prediction of onset of labor. *Am. J. Obstet. Gynecol.*, 1967; 99: 546–550.

144. Bauer F., Wujanz G. Ergebnisse der Geburtseinleitung am "biologischen" Geburtstermin durch tiefe Blasensprengung. *Geburtshilfe und Frauenheilkunde*, 1967; 8: 798–800.

145. Benson R.C. Poloznictwo i ginekologia. *PZWL*, 1988.

146. Birnholz J.C. Ecologic physiology of the fetus. *Radiol. Clin. North Am.*, 1990; 28: 179–188.

147. Campbell S. Jan Donald's child comes of age. *Ultrasound Obstet. Gynecol.*, 1991; 1: 1.

148. Chervenak F.A., Berkowitz G.S. *et al.* A comparison of sonographic estimation of fetal weight and obstetrically determined gestational age in the prediction of neonatal outcome for the very low-birth weight fetus. *Am. J. Obstet. Gynecol.*, 1985; 152: 47.

149. Eik-Nes S.H., Okland O., Aure J.C. *et al.* Ultrasound screening in pregnacy. *Lancet*, 1984; 1: 1347.

150. Llewellyn-Jones D. *Fundamentals of Obstetrics and Gynecology*. Farber and Farber, London and Boston, 1982.

151. Maeda K., Hogaki M., Nakano H. (eds.) *Computers and Perinatal Medicine*. Excerpta Medica, Amsterdam, 1990.

152. Ohel G., Granat M., Zeevi D., Golan A. *et al.* Advanced ultrasonic placental maturation in twin pregnancies. *Am. J. Obstet. Gynecol.*, 1987; 156: 76.

153. Okonofua F.E., Atoyebi F.A. Accuracy of prediction of gestational age by ultrasound measurement of biparietal diameter in Nigerian women. Int. J. Gynecol. Obstet., 1989; 28: 217–219.

154. Queenan J.T. Antenatal detection of congenital malformation. In: Klimek R. (ed.) *Pre- and Perinatal Psycho-medicine*. DWN DReAM. Cracow, 1992, p. 123–133.

155. Sabbagha R.E. Assessment of gestational age. In: *Materials of Postgraduate Course in Management of High Risk Pregnancy*. Cracow, Poland, 1984. Project HOPE

156. Sabbagha R.E., Minogue J., Tamura R.K., Hungerford S.A. Estimation of birth weight by use of ultrasonographic formulas targeted to large-, appropriate-, and small-for-gestational-age fetuses. *Am. J. Obstet. Gynecol.*, 1989; 160: 854.

157. Satin A.J., Hankins G.D.V. *et al.* A randomized study of two dosing regimens of oxytocin for the induction of patients with an unfavorable cervix. *Am. J. Obstet. Gynecol.*, 1991; 164: 1-Part 2: 307.

158. Schenker J.G., Weinstein D. (eds.) *The Intrauterine Life of Fetus*. Excerpta Medica, 1986.

159. Schwarz S., Klimek R. Enzymatic studies on the development of pregnancy and onset of labor. XV Zjazd PTG Gdansk 1962. *Pol. Med. Sci. Hist. Bull.*, 1964; 7: 77.

160. Stanek J. Enzymatyczno-cytohormonalne monitorowanie zagrozonej ciazy. Rozprawa doktorska AM Kraków, 1974.

161. Stanek J. Enzymatyczno-cytohormonalne monitorowanie endokrynologicznie zagrozonej ciazy. *Folia Med. Cracoviensia*, 1975; 4: 365–379.

162. Stanek J. Colpocytograms and maternal serum placental cystine aminopeptidase, tissue cystine aminopeptidase, alkaline phosphatase and heat stable alkaline phosphatase activity in monitoring the last four weeks before delivery in high-risk pregnancy. *Acta Cytologica*, 1977; 21: 229–235.

163. Stanek J. Ilosciowa ocena kolpocytogramów w ciazy o niepowiklanym przebiegu. *Folia Med. Cracoviensia*, 1979; 21: 96–116.

164. Vintrileos A.M., Campbell W.A. *et al.* Fetal weight estimation formulas with head, abdominal, femur. *Am. J. Obstet. Gynecol.*, 1987; 157: 410.

165. Waldenstrom V., Axelsson O., Nisson S. *et al.* Effects of routine one-stage ultrasound screening. *Lancet*, 1988; 2: 585.

166. Wharton B.A., Dunn P.M. (eds.) Perinatal growth. *Acta Paediatr. Scand.*, Suppl., 1985.

167. Wolf H., Oosting H. Placental volume measurement by ultrasonography: evaluation of the method. *Am. J. Obstet. Gynecol.*, 1987; 156: 1191.

168. Zidovsky J. Ein Vergleich röntgenologischer und zytodiagnosticher Bestimmungsmöglichkeiten des Geburtstermins und der Fruchtschüdigung. *Zentralblatt Gynakol.*, 1964; 86: 588.

169. Ryden G. Cystine aminopeptidase activity in pregnancy. *Acta Obstet. Gynecol. Scand.*, 1972; 51: 329–334.

170. Tomaszczyk J., Balajewicz M., Tomaszczyk B. Przedporodowy profil oksytocynazemii u matki a stan noworodka. *Gin. Pol.*, 1984; 9: 649.

171. Klimek R. The synergy of perinatal medicine and psychology. *Int. J. Prenatal and Perinatal Stud.*, 1989; 1: 251.

172. Magin R.L., Liburdy R.P., Persson B. (eds.) Biological effects and safety aspects of NMRI-S. *Ann NIAS*, 1992, v. 649.

173. Stehling M.K., Turner R., Mansfield P. Echo-planar imaging: magnetic resonance imaging in a fraction of a second. *Science*, 1991; 254: 43.

174. Ferguson R., Stephen A.M. The effect of race on relationship between fetal death and altered fetal growth. *Am. J. Obstet. Gynecol.*, 1990; 163: 1222–1230.

175. Klimek R. Estimation of biological gestational age by corrected Ballard maturity rating. *Int. J. Prenatal Perinatal Stud.*, 1992, 4; Suppl. 1, 102.

176. Klimek R. Psychomedicine. In: Klimek R. (ed.) *Pre- and Perinatal Psychomedicine.* DWN DReAM Cracow, 1992.

176. Zador I.E., Salari V., Chik L., Sokol R.J. Ultrasound measurement of the fetal head: computer vs. operator. *Ultrasound Obstet. Gynecol.*, 1991; 1: 208.

177. Lange R.C., Duberg A.C., McCarthy S.M. An evaluation of MRI contrast in the uterus using synthetic imaging. *Magnetic Resonance Med.*, 1991; 17: 279–284.

178. Šorm F. *et al.* In: *Oxytocin and Its Analogues.* Klimek R., Król W. (eds.) Cracow PTE, 1964.

179. Geirsson R.T. Ultrasound instead of LMP as the basis of gestational age assignment. *Ultrasound Obstet. Gynecol.*, 1991; 1: 212.

180. Klimek R., Klimek M. Biological gestational age and its calendar assessment with ultrasound. Part I and Part II. *Gynäkol. Geburtsh. Rundschau*, 1992; 32: 159–163.

181. Sciarra J.J. Obstetrics and gynecology: the next ten years. In: Klimek R. (ed.) *Pre- and Perinatal Psycho-medicine.* DWN DReAM. Cracow, 1992, p. 37–47.

182. Skret A., Klimek R., Cebulak K., Klimek M., Palczak R., Lassota L., Dorski A., Janeczko J. Flowmetric biological age in advanced pregnancy. *Int. J. Prenatal Perinatal Stud.*, 1992. 4; Suppl. 1, 123.

183. Slomko Z., red. Medycyna Perinatalna. *PZWL*, Warszawa, 1985–I; 1986–II.

184. Van Oudheusden A.P.M. *Klinisch Chemich Onderzoek over de Activiteitsmeting van Cystine-aminopeptidase (Oxytocinase).* Drukkerij Elinkwijk B.V., Utrecht, 1973.

187. Teter J., red. Leki hormonalne – klinika i terapia. *PZWL*, Warszawa, 1979.

Index